Even You Can Learn Statistics and Analytics

Third Edition

An Easy to Understand Guide to Statistics and Analytics

David M. Levine, Ph.D.

David F. Stephan

Editor-in-Chief: Amy Neidlinger
Operations Specialist: Jodi Kemper
Cover Designer: Alan Clements
Managing Editor: Kristy Hart
Senior Project Editor: Betsy Gratner
Copy Editor: Krista Hansing
Proofreader: Sarah Kearns
Interior Designer: Argosy
Compositor: codeMantra
Manufacturing Buyer: Dan Uhrig

For information about buying this title in bulk quantities, or for special sales opportunities (which may include electronic versions; custom cover designs; and content particular to your business, training goals, marketing focus, or branding interests), please contact our corporate sales department at corpsales@pearsoned.com or (800) 382-3419.

For government sales inquiries, please contact governmentsales@pearsoned.com.

For questions about sales outside the U.S., please contact international@pearsoned.com.

Printed in the United States of America

First Printing December 2014

ISBN-10: 0-13-338266-4
ISBN-13: 978-0-13-338266-2

Pearson Education LTD.
Pearson Education Australia PTY, Limited.
Pearson Education Singapore, Pte. Ltd.
Pearson Education North Asia, Ltd.
Pearson Education Canada, Ltd.
Pearson Educación de Mexico, S.A. de C.V.
Pearson Education—Japan
Pearson Education Malaysia, Pte. Ltd.

Library of Congress Control Number: 2014949421

To our wives
Marilyn and Mary

To our children
Sharyn and Mark

And to our parents
In loving memory, Lee, Reuben, Ruth, and Francis

Table of Contents

Acknowledgments

We would especially like to thank the staff at Financial Times/Pearson: Amy Neidlinger for making this book a reality, Sarah Kearns for her proofreading, Krista Hansing for her copy editing, and Betsy Gratner for her work in the production of this text.

We have sought to make the contents of this book as clear, accurate, and error-free as possible. We invite you to make suggestions or ask questions about the content if you think we have fallen short of our goals in any way. Please email your comments to davidlevine@davidlevinestatistics.com and include "Even You Can Learn Statistics and Analytics 3/e" in the subject line.

About the Authors

David M. Levine is Professor Emeritus of Statistics and Computer Information Systems at Baruch College-CUNY. He received B.B.A. and M.B.A. degrees in Statistics from City College of New York and a Ph.D. degree from New York University in Industrial Engineering and Operations Research. He is nationally recognized as a leading innovator in business statistics education and is the coauthor of such best-selling statistics textbooks as *Statistics for Managers Using Microsoft Excel*, *Basic Business Statistics: Concepts and Applications*, *Business Statistics: A First Course*, and *Applied Statistics for Engineers and Scientists Using Microsoft Excel and Minitab*.

He also is the author of *Statistics for Six Sigma Green Belts and Champions*, published by Financial Times–Prentice-Hall. He is coauthor of *Six Sigma for Green Belts and Champions* and *Design for Six Sigma for Green Belts and Champions* also published by Financial Times–Prentice-Hall, and *Quality Management* Third Ed., McGraw-Hill-Irwin. He is also the author of *Video Review of Statistics* and *Video Review of Probability*, both published by Video Aided Instruction. He has published articles in various journals including *Psychometrika*, *The American Statistician*, *Communications in Statistics*, *Multivariate Behavioral Research*, *Journal of Systems Management*, *Quality Progress*, and *The American Anthropologist* and has given numerous talks at American Statistical Association, Decision Sciences Institute, and Making Statistics More Effective in Schools of Business conferences. While at Baruch College, Dr. Levine received numerous awards for outstanding teaching.

David F. Stephan is an independent instructional technologist. During his more than 20 years teaching at Baruch College-CUNY, he pioneered the use of computer-equipped classrooms and interdisciplinary multimedia tools, was an associate director of a U.S. Department of Education FIPSE project that applied interactive media to support instruction and devised techniques for teaching computer applications in a business context. A frequent participant in the Decision Sciences Institute's Making Statistics for More Effective in School of Business mini-conferences, he is also a coauthor of *Business Statistics: A First Course* and *Statistics for Managers Using Microsoft Excel*. He is also the developer of PHStat, the statistics add-in for Microsoft Excel distributed by Pearson Education.

About the Authors

Introduction

The *Even You Can Learn Statistics and Analytics* Owners Manual

In today's world, understanding statistics and analytics is more important than ever. *Even You Can Learn Statistics and Analytics: A Guide for Everyone Who Has Ever Been Afraid of Statistics and Analytics* can teach you the basic concepts that provide you with the knowledge to apply statistics and analytics in your life. You will also learn the most commonly used statistical methods and have the opportunity to practice those methods while using the Microsoft Excel spreadsheet program.

Please read the rest of this introduction so that you can become familiar with the distinctive features of this book. You can also visit the website for this book (**www.ftpress.com/evenyoucanlearnstatistics3e**) where you can learn more about this book as well as download files that support your learning of statistics.

Mathematics Is Always Optional!

Never mastered higher mathematics—or generally fearful of math? Not to worry, because in *Even You Can Learn Statistics and Analytics,* you will find that every concept is explained in plain English, without the use of higher mathematics or mathematical symbols. Interested in the mathematical foundations behind statistics? *Even You Can Learn Statistics and Analytics* includes **Equation Blackboards**, stand-alone sections that present the equations behind statistical methods and complement the main material. Either way, you can learn statistics.

Learning with the Concept-Interpretation Approach

Even You Can Learn Statistics and Analytics uses a **Concept-Interpretation** approach to help you learn statistics. For each important statistical concept, you will find the following:

- A **CONCEPT**, a plain language definition that uses no complicated mathematical terms.
- An **INTERPRETATION**, that fully explains the concept and its importance to statistics. When necessary, these sections also discuss common

misconceptions about the concept as well as the common errors people can make when trying to apply the concept.

For simpler concepts, an **EXAMPLES** section lists real-life examples or applications of the statistical concepts. For more involved concepts, **WORKED-OUT PROBLEMS** provide a complete solution to a statistical problem—including actual spreadsheet and calculator results—that illustrate how you can apply the concept to your own situations.

Practicing Statistics While You Learn Statistics

To help you learn statistics, you should always review the worked-out problems that appear in this book. As you review them, you can practice what you have just learned by using the optional **SPREADSHEET SOLUTION** sections.

Spreadsheet Solution sections enable you to use Microsoft Excel as you learn statistics.

Prefer to practice using a calculator? **CALCULATOR KEYS** sections (available online at **www.ftpress.com/evenyoucanlearnstatistics3e**) provide you with the step-by-step instructions to perform statistical analysis using one of the calculators from the Texas Instruments TI-83/84 family. (You can adapt many instruction sets for use with other TI statistical calculators.)

If you don't want to practice your spreadsheet skills, you can examine the spreadsheet results that appear throughout the book (or the calculator results available online). Many spreadsheet results are available as files that you can download for free at **www.ftpress.com/evenyoucanlearnstatistics3e**.

Spreadsheet program users will also benefit from Appendix D and Appendix E, which help teach you more about spreadsheets as you learn statistics.

And if technical issues or instructions have ever confounded your using Microsoft Excel in the past, check out Appendix A, which details the technical configuration issues you might face and explains the conventions used in all technical instructions that appear in this book.

In-Chapter Aids

As you read a chapter, look for the following icons for extra help:

Important Point icons highlight key definitions and explanations.

File icons identify files that allow you to examine the data in selected problems. (You can download these files for free at **www.ftpress.com/evenyoucanlearnstatistics3e.**)

Interested in the mathematical foundations of statistics? Then look for the Interested in Math? icons throughout the book. But remember, you can skip any or all of the math sections without losing any comprehension of the statistical methods presented, because math is always optional in this book!

End-of-Chapter Features

At the end of most chapters of *Even You Can Learn Statistics and Analytics,* you can find the following features, which you can review to reinforce your learning.

Important Equations

The **Important Equations** sections present all of the important equations discussed in the chapter. You can use these lists for reference and later study even if you have skipped over the Equation Blackboards and "interested in math" passages.

One-Minute Summaries

One-Minute Summaries are a quick review of the significant topics of a chapter in outline form. When appropriate, the summaries also help guide you to make the right decisions about applying statistics to the data you seek to analyze.

Test Yourself

The **Test Yourself** sections offer a set of short-answer questions and problems that enable you to review and test yourself (with answers provided) to see how much you have retained of the concepts presented in a chapter.

New to the Third Edition

The third edition of this book includes these features, which earlier editions did not contain:

- A new chapter that introduces the concepts and application of analytics, a new and growing part of statistics (Chapter 12)
- New chapters that illustrate descriptive and predictive analytical methods (Chapters 13 and 14)
- A new and expanded discussion about using Microsoft Excel, focused on using Excel 2013 (Microsoft Windows), Excel 2011 (OS X), and Office 365 Excel

The book also contains many revised in-chapter, worked-out problems and many new or revised end-of-chapter problems.

Summary

Even You Can Learn Statistics and Analytics can help you whether you are studying statistics as part of a formal course, just brushing up on your knowledge of statistics for a specific analysis, or need to learn about analytics. Be sure to visit the website for this book (**www.ftpress.com/evenyoucanlearnstatistics3e**) and feel free to contact the authors via email at davidlevine@davidlevinestatistics.com; include *Even You Can Learn Statistics and Analytics 3/e* in the subject line if you have any questions about this book.

Fundamentals of Statistics

Every day, the media uses numbers to describe or analyze our world:

- **"6 New Facts About Facebook"** (A. Smith, **www.pewresearch.org/author/asmith**, 3 February 2014). A survey reported that women were more likely than men to cite seeing photos or videos, sharing with many people at once, seeing entertaining or funny posts, learning about ways to help others, and receiving support from people in their network as reasons to use Facebook.
- **"First Two Years of College Wasted?"** (M. Marklein, *USA Today*, 18 January 2011, p. 3A). A survey of more than 3,000 full-time, traditional-age students found that the students spent 51% of their time on socializing, recreation, and other activities; 9% of their time attending class and labs; and 7% of their time studying.
- **"Follow the Tweets"** (H. Rui, A. Whinston, and E. Winkler, *The Wall Street Journal*, 30 November 2009, p. R4). In this study, the authors found that the number of times a specific product was mentioned in comments in the Twitter social messaging service could be used to make accurate predictions of sales trends for that product.

You can make better sense of the numbers you encounter if you learn to understand statistics. **Statistics**, a branch of mathematics, uses procedures that allow you to correctly analyze the numbers. These procedures, or **statistical methods**, transform numbers into useful information that you can use when making decisions about the numbers. Statistical methods can also tell you the known risks associated with making a decision as well as help you make more consistent judgments about the numbers.

Learning statistics requires you to reflect on the significance and the importance of the results to the decision-making process you face. This statistical interpretation means knowing when to ignore results because they are misleading, are produced by incorrect methods, or just restate the obvious, as in "100% of the authors of this book are named 'David.'"

In this chapter, you begin by learning five basic words—*population*, *sample*, *variable*, *parameter*, and *statistic* (singular)—that identify the fundamental concepts of statistics. These five words, and the other concepts introduced in this chapter, help you explore and explain the statistical methods discussed in later chapters.

1.1 The First Three Words of Statistics

You've already learned that statistics is about analyzing things. Although *numbers* was the word used to represent things in the opening of this chapter, the first three words of statistics, *population*, *sample*, and *variable*, help you to better identify what you analyze with statistics.

Population

CONCEPT All the members of a group about which you want to reach a conclusion.

EXAMPLES All U.S. citizens who are currently registered to vote, all patients treated at a particular hospital last year, the entire set of individuals who accessed a website on a particular day.

Sample

CONCEPT The part of the population selected for analysis.

EXAMPLES The registered voters selected to participate in a recent survey concerning their intention to vote in the next election, the patients selected to fill out a patient satisfaction questionnaire, 100 boxes of cereal selected from a factory's production line, 500 individuals who accessed a website on a particular day.

Variable

CONCEPT A characteristic of an item or an individual that will be analyzed using statistics.

EXAMPLES Gender, the party affiliation of a registered voter, the household income of the citizens who live in a specific geographical area, the publishing category (hardcover, trade paperback, mass-market paperback, textbook) of a book, the number of cell phones in a household.

INTERPRETATION All the variables taken together form the data of an analysis. Although people often say that they are analyzing their data, they are, more precisely, analyzing their variables.

You should distinguish between a variable, such as gender, and its value for an individual, such as male. An **observation** is all the values for an individual item in the sample. For example, a survey might contain two variables, gender and age. The first observation might be male, 40. The second observation might be female, 45. The third observation might be female, 55. By convention, when you organize data in tabular form, you place the values for a variable to be analyzed in a column. Therefore, some people refer to a variable as a *column of data*. Likewise, some people call an observation a *row of data*.

	Categorical Variables	Numerical Variables
Concept	The values of these variables are selected from an established list of categories.	The values of these variables involve a counted or measured value.
Subtypes	None	**Discrete** values are counts of things. **Continuous** values are measures and any value can theoretically occur, limited only by the precision of the measuring process.
Examples	Gender, a variable that has the categories "male" and "female." variable. Academic major, a variable that might have the categories "English," "Math," "Science," and "History," among others.	The number of people living in a household, a discrete numerical variable. The time it takes for someone to commute to work, a continuous variable.

important point

All variables should have an operational definition—that is, a universally accepted meaning that is understood by all associated with an analysis. Without operational definitions, confusion can occur. A famous example of such confusion was a survey that asked about *sex* and a number of survey takers answered yes and not male or female, as the survey writer intended.

1.2 The Fourth and Fifth Words

After you know what you are analyzing, or, using the words of Section 1.1, after you have identified the variables from the population or sample under study, you can define the **parameters** and **statistics** that your analysis will determine.

Parameter

CONCEPT A numerical measure that describes a variable (characteristic) from a population.

EXAMPLES The percentage of all registered voters who intend to vote in the next election, the percentage of all patients who are very satisfied with the care they received, the mean time that all visitors spent on a website during a particular day.

Statistic

CONCEPT A numerical measure that describes a variable (characteristic) of a sample (part of a population).

EXAMPLES The percentage of registered voters in a sample who intend to vote in the next election, the percentage of patients in a sample who are very satisfied with the care they received, the mean time that a sample of visitors spent on a website during a particular day.

INTERPRETATION Calculating statistics for a sample is the most common activity because collecting population data is impractical in many actual decision-making situations.

1.3 The Branches of Statistics

You can use parameters and statistics either to describe your variables or to reach conclusions about your data. These two uses define the two branches of statistics: **descriptive statistics** and **inferential statistics**.

Descriptive Statistics

CONCEPT The branch of statistics that focuses on collecting, summarizing, and presenting a set of data.

EXAMPLES The mean age of citizens who live in a certain geographical area, the mean length of all books about statistics, the variation in the time that visitors spent visiting a website.

INTERPRETATION You are most likely to be familiar with this branch of statistics because many examples arise in everyday life. Descriptive statistics serves as the basis for analysis and discussion in fields as diverse as securities trading, the social sciences, government, the health sciences, and professional sports. Descriptive methods can seem deceptively easy to apply because they are often easily accessible in calculating and computing devices. However, this ease does not mean that descriptive methods are without their pitfalls, as Chapter 2 and Chapter 3 explain.

Inferential Statistics

CONCEPT The branch of statistics that analyzes sample data to reach conclusions about a population.

EXAMPLE A survey that sampled 1,264 women found that 45% of those polled considered friends or family as their most trusted shopping advisers and only 7% considered advertising as their most trusted shopping adviser. By using methods discussed in Section 6.4, you can use these statistics to draw conclusions about the population of all women.

INTERPRETATION When you use inferential statistics, you start with a hypothesis and look to see whether the data are consistent with that hypothesis. This deeper level of analysis means that inferential statistical methods can be easily misapplied or misconstrued, and that many inferential methods require a calculating or computing device. (Chapters 6 through 9 discuss some of the inferential methods that you will most commonly encounter.)

1.4 Sources of Data

You begin every statistical analysis by identifying the source of the data that you will use for **data collection**. Among the important sources of data are **published sources**, **experiments**, and **surveys**.

Published Sources

CONCEPT Data available in print or in electronic form, including data found on Internet websites. Primary data sources are those published by the individual or group that collected the data. Secondary data sources are those compiled from primary sources.

EXAMPLE Many U.S. federal agencies, including the Census Bureau, publish primary data sources that are available at the **www.fedstats.gov** website. Industry-specific groups and business news organizations commonly publish online or in-print secondary source data compiled by business organizations and government agencies.

INTERPRETATION You should always consider the possible bias of the publisher and whether the data contain all the necessary and relevant variables when using published sources. This is especially true of sources found through Internet search engines.

Experiments

CONCEPT A study that examines the effect on a variable of varying the value(s) of another variable or variables, while keeping all other things equal. A typical experiment contains both a treatment group and a control group. The treatment group consists of those individuals or things that receive the treatment(s) being studied. The control group consists of those individuals or things that do not receive the treatment(s) being studied.

EXAMPLE Pharmaceutical companies use experiments to determine whether a new drug is effective. A group of patients who have many similar characteristics is divided into two subgroups. Members of one group, the treatment group, receive the new drug. Members of the other group, the control group, often receive a placebo, a substance that has no medical effect. After a time period, statistics about each group are compared.

INTERPRETATION Proper experiments are either single-blind or double-blind. A study is a single-blind experiment if only the researcher conducting the study knows the identities of the members of the treatment and control groups. If neither the researcher nor study participants know who is in the treatment group and who is in the control group, the study is a double-blind experiment.

When conducting experiments that involve placebos, researchers also have to consider the placebo effect—that is, whether people in the control group will improve because they believe they are getting a real substance that is intended to produce a positive result. When a control group shows as much improvement as the treatment group, a researcher can conclude that the placebo effect is a significant factor in the improvements of both groups.

Surveys

CONCEPT A process that uses questionnaires or similar means to gather values for the responses from a set of participants.

EXAMPLES The decennial U.S. census mail-in form, a poll of likely voters, a website instant poll or "question of the day."

INTERPRETATION Surveys are either **informal**, open to anyone who wants to participate; **targeted**, directed toward a specific group of individuals; or include people chosen at random. The type of survey affects how the data collected can be used and interpreted.

1.5 Sampling Concepts

In the definition of **statistic** in Section 1.2, you learned that calculating statistics for a sample is the most common activity because collecting population data is usually impractical. Because samples are so commonly used, you need to learn the concepts that help identify all the members of a population and that describe how samples are formed.

Frame

CONCEPT The list of all items in the population from which the sample will be selected.

EXAMPLES Voter registration lists, municipal real estate records, customer or human resource databases, directories.

INTERPRETATION Frames influence the results of an analysis, and using different frames can lead to different conclusions. You should always be careful to make sure your frame completely represents a population; otherwise, any sample selected will be biased, and the results generated by analyses of that sample will be inaccurate.

Sampling

CONCEPT The process by which members of a *population* are selected for a *sample*.

EXAMPLES Choosing every fifth voter who leaves a polling place to interview, selecting playing cards randomly from a deck, polling every tenth visitor who views a certain website today.

INTERPRETATION Some sampling techniques, such as an "instant poll" found on a web page, are naturally suspect as such techniques do not depend on a well-defined frame. The sampling technique that uses a well-defined frame is **probability sampling**.

Probability Sampling

CONCEPT A sampling process that considers the chance of selection of each item. Probability sampling increases your chance that the sample will be representative of the population.

EXAMPLES The registered voters selected to participate in a recent survey concerning their intention to vote in the next election, the patients selected to fill out a patient-satisfaction questionnaire, 100 boxes of cereal selected from a factory's production line.

INTERPRETATION You should use probability sampling whenever possible, because *only* this type of sampling enables you to apply inferential statistical methods to the data you collect. In contrast, you should use nonprobability sampling, in which the chance of occurrence of each item being selected is not known, to obtain rough approximations of results at low cost or for small-scale, initial, or pilot studies that will later be followed up by a more rigorous analysis. Surveys and polls that invite the public to call in or answer questions on a web page are examples of nonprobability sampling.

Simple Random Sampling

CONCEPT The probability sampling process in which every individual or item from a population has the same chance of selection as every other individual or item. Every possible sample of a certain size has the same chance of being selected as every other sample of that size.

EXAMPLES Selecting a playing card from a shuffled deck or using a statistical device such as a table of random numbers.

INTERPRETATION Simple random sampling forms the basis for other random sampling techniques. The word *random* in this phrase requires clarification. In this phrase, *random* means no repeating patterns—that is, in a given sequence, a given pattern is equally likely (or unlikely). It does not refer to the most commonly used meaning of "unexpected" or "unanticipated" (as in "random acts of kindness").

Other Probability Sampling Methods

Other, more complex, sampling methods are also used in survey sampling. In a stratified sample, the items in the frame are first subdivided into separate subpopulations, or strata, and a simple random sample is selected within each of the strata. In a cluster sample, the items in the frame are divided into several clusters so that each cluster is representative of the entire population. A random sampling of clusters is then taken, and all the items in each selected cluster or a sample from each cluster are then studied.

1.6 Sample Selection Methods

Sampling can be done either with or without replacement of the items being selected. Almost all survey sampling is done without replacement.

Sampling with Replacement

CONCEPT A sampling method in which each selected item is returned to the frame from which it was selected so that it has the same probability of being selected again.

EXAMPLE Selecting items from a fishbowl and returning each item to it after the selection is made.

Sampling Without Replacement

CONCEPT A sampling method in which each selected item is not returned to the frame from which it was selected. Using this technique, an item can be selected no more than one time.

EXAMPLES Selecting numbers in state lottery games, selecting cards from a deck of cards during games of chance such as blackjack or poker.

INTERPRETATION Sampling without replacement means that an item can be selected no more than one time. You should choose sampling without replacement instead of sampling with replacement because statisticians generally consider the former to produce more desirable samples.

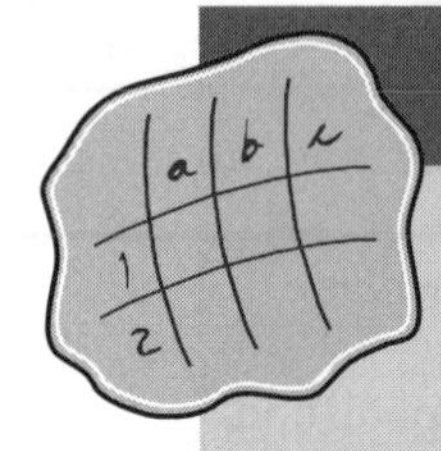

spreadsheet solution

Entering Data

Enter the data values of a variable in a blank column of a worksheet. Use the row 1 cell for the variable name.

To create a new file (workbook) that contains a blank worksheet for your entries, select **File** → **New** and, in the New panel, double-click the **Blank Workbook** icon. (In Excel 2007, click the **Office Button** instead of selecting **File**.) To enter data into a specific cell, move the cell pointer to that cell by using the cursor keys, moving the mouse pointer, or completing the proper touch operation. As you type an entry, the entry appears in the formula bar area located over the top of the worksheet. You complete your entry by pressing **Tab** or **Enter** or by clicking the checkmark button in the formula bar.

To save your new file, select **File** → **Save As** and, in the Save As dialog box, navigate to the folder where you want to save your file. Accept or revise the filename and then click **Save**. To later retrieve the file, select **File** → **Open** and in the Open dialog box, navigate to the folder that contains the desired file, select the desired file from the list, and then click **Open**. (In Excel 2007, you begin these operations by clicking the **Office Button**, not by selecting **File**.)

Throughout this book, the symbol → links a sequence of Ribbon or menu selections. **File** → **New** means to first select the File tab and then select New from the list that appears.

One-Minute Summary

Mastering basic vocabulary is the first step in learning statistics. Understanding the types of statistical methods, the sources of data used for data collection, sampling methods, and the types of variables used in statistical analysis are also important introductory concepts. Subsequent chapters focus on four important reasons for learning statistics:

- To present and describe information (Chapters 2 and 3)
- To reach conclusions about populations based only on sample results (Chapters 4 through 9)
- To develop reliable forecasts (Chapters 10 and 11)
- To use analytics to reach conclusions about large sets of data (Chapters 12, 13, 14)

Test Yourself

1. The portion of the population that is selected for analysis is called:
   (a) a sample
   (b) a frame
   (c) a parameter
   (d) a statistic

2. A summary measure that is computed from only a sample of the population is called:
   (a) a parameter
   (b) a population
   (c) a discrete variable
   (d) a statistic

3. The height of an individual is an example of a:
   (a) discrete variable
   (b) continuous variable
   (c) categorical variable
   (d) constant

4. The body style of an automobile (sedan, minivan, SUV, and so on) is an example of a:
   (a) discrete variable
   (b) continuous variable
   (c) categorical variable
   (d) constant

5. The number of credit cards in a person's wallet is an example of a:
   (a) discrete variable
   (b) continuous variable
   (c) categorical variable
   (d) constant

6. Statistical inference occurs when you:
   (a) compute descriptive statistics from a sample
   (b) take a complete census of a population
   (c) present a graph of data
   (d) take the results of a sample and reach conclusions about a population

7. The human resources director of a large corporation wants to develop a dental benefits package and decides to select 100 employees from a list of all 5,000 workers in order to study their preferences for the various components of a potential package. All the employees in the corporation constitute the _______.
   (a) sample
   (b) population
   (c) statistic
   (d) parameter

8. The human resources director of a large corporation wants to develop a dental benefits package and decides to select 100 employees from a list of all 5,000 workers in order to study their preferences for the various components of a potential package. The 100 employees who will participate in this study constitute the _______.
   (a) sample
   (b) population
   (c) statistic
   (d) parameter

9. Those methods that involve collecting, presenting, and computing characteristics of a set of data in order to properly describe the various features of the data are called:
   (a) statistical inference
   (b) the scientific method
   (c) sampling
   (d) descriptive statistics

10. Based on the results of a poll of 500 registered voters, the conclusion that the Democratic candidate for U.S. president will win the upcoming election is an example of:
   (a) inferential statistics
   (b) descriptive statistics
   (c) a parameter
   (d) a statistic

11. A numerical measure that is computed to describe a characteristic of an entire population is called:
   (a) a parameter
   (b) a population
   (c) a discrete variable
   (d) a statistic

12. You were working on a project to examine the value of the American dollar as compared to the English pound. You accessed an Internet site where you obtained this information for the past 50 years. Which method of data collection were you using?
    (a) published sources
    (b) experimentation
    (c) surveying

13. Which of the following is a discrete variable?
    (a) The favorite flavor of ice cream of students at your local elementary school
    (b) The time it takes for a certain student to walk to your local elementary school
    (c) The distance between the home of a certain student and the local elementary school
    (d) The number of teachers employed at your local elementary school

14. Which of the following is a continuous variable?
    (a) The eye color of children eating at a fast-food chain
    (b) The number of employees of a branch of a fast-food chain
    (c) The temperature at which a hamburger is cooked at a branch of a fast-food chain
    (d) The number of hamburgers sold in a day at a branch of a fast-food chain

15. The number of cell phones in a household is an example of:
    (a) a categorical variable
    (b) a discrete variable
    (c) a continuous variable
    (d) a statistic

Answer True or False:

16. The possible responses to the question, "How long have you been living at your current residence?" are values from a continuous variable.
17. The possible responses to the question, "How many times in the past seven days have you streamed a movie or TV show online?" are values from a discrete variable.

Fill in the blank:

18. An insurance company evaluates many variables about a person before deciding on an appropriate rate for automobile insurance. The number of accidents a person has had in the past three years is an example of a _______ variable.

19. An insurance company evaluates many variables about a person before deciding on an appropriate rate for automobile insurance. The distance a person drives in a day is an example of a _______ variable.
20. An insurance company evaluates many variables about a person before deciding on an appropriate rate for automobile insurance. A person's marital status is an example of a _______ variable.
21. A numerical measure that is computed from only a sample of the population is called a ______.
22. The portion of the population that is selected for analysis is called the ______.
23. A college admission application includes many variables. The number of advanced placement courses the student has taken is an example of a ______ variable.
24. A college admission application includes many variables. The gender of the student is an example of a ______ variable.
25. A college admission application includes many variables. The distance from the student's home to the college is an example of a ______ variable.

Answers to Test Yourself

1. a
2. d
3. b
4. c
5. a
6. d
7. b
8. a
9. d
10. a
11. a
12. a
13. d
14. c
15. b
16. True
17. True
18. discrete
19. continuous
20. categorical
21. statistic
22. sample
23. discrete
24. categorical
25. continuous

References

1. Berenson, M. L., D. M. Levine, and K. A. Szabat. *Basic Business Statistics: Concepts and Applications*, Thirteenth Edition. Upper Saddle River, NJ: Pearson Education, 2015.
2. Cochran, W. G. *Sampling Techniques*, Third Edition. New York: John Wiley & Sons, 1977.

CHAPTER

2

Presenting Data in Tables and Charts

Tables and charts are ways of summarizing categorical and numerical variables that can help you present information effectively. In this chapter, you will learn the appropriate types of tables and charts to use for each type of variable.

2.1 Presenting Categorical Variables

You present a categorical variable by first sorting values according to the categories of the variable. Then you place the count, amount, or percentage (part of the whole) of each category into a summary table or into one of several types of charts.

The Summary Table

CONCEPT A two-column table in which category names are listed in the first column and the count, amount, or percentage of values are listed in a second column. Sometimes, additional columns present the same data in more than one way (for example, as counts and percentages).

EXAMPLE The results of a survey that asked young adults about the main reason they shop online can be presented using a summary table:

Reason	Percentage
Better prices	37%
Avoiding holiday shopping	29%
Convenience	18%
Better selection	13%
Ships directly	3%

Source: Data extracted and adapted from "Main Reason Young Adults Shop Online?," *USA Today*, December 5, 2012, p. 1A.

INTERPRETATION Summary tables enable you to see the big picture about a set of data. In this example, you can conclude that 37% shop online mainly for better prices and convenience and that 29% shop online mainly to avoid holiday shopping.

The Bar Chart

CONCEPT A chart containing rectangles ("bars") in which the length of each bar represents the count, amount, or percentage of responses of one category.

EXAMPLE This percentage bar chart presents the data of the summary table discussed in the previous example:

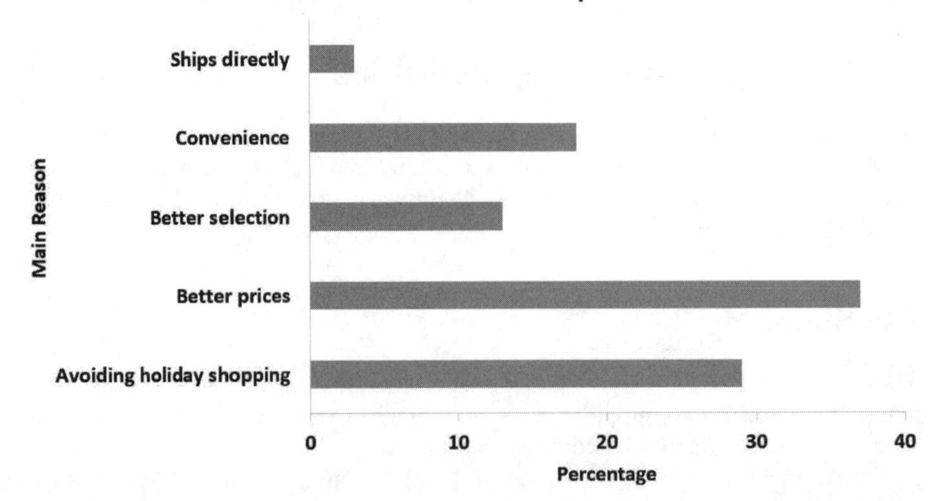

INTERPRETATION A bar chart is better than a summary table at making the point that the category better prices is the single largest category for this example. For most people, scanning a bar chart is easier than scanning a column of numbers in which the numbers are unordered, as they are in the bill payment summary table.

The Pie Chart

CONCEPT A circle chart in which wedge-shaped areas—pie slices—represent the count, amount, or percentage of each category and the entire circle ("pie") represents the total.

EXAMPLE This pie chart presents the data of the summary table discussed in the preceding two examples:

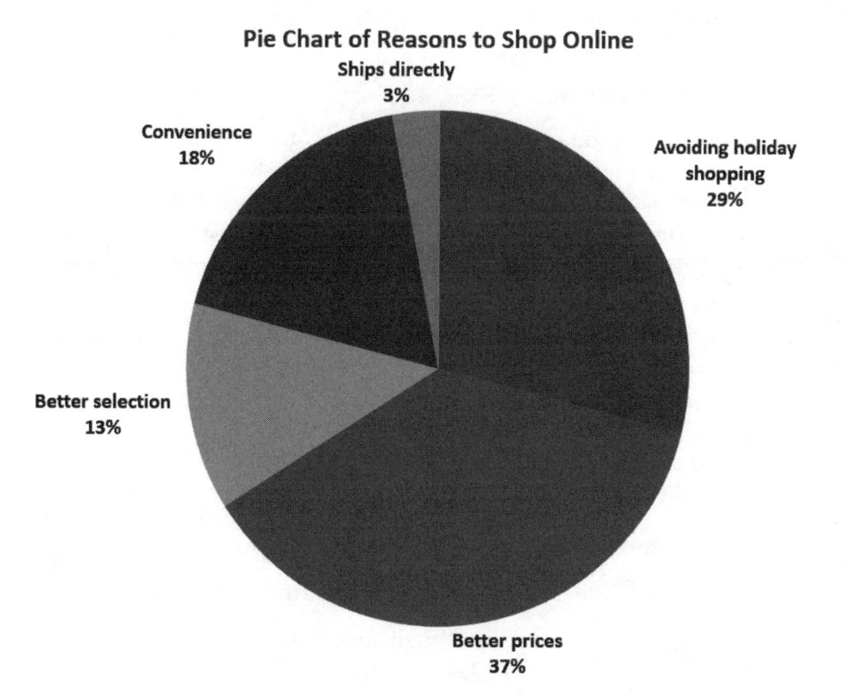

INTERPRETATION The pie chart enables you to see each category's portion of the whole. You can see that more young adults shopped online for better prices or to avoid holiday shopping, a small number shopped online for better selection, and that hardly anyone shopped online because of direct shipment.

Besides using programs such as Microsoft Excel to create a pie chart, you can also create a pie chart using a protractor to divide up a hand-drawn circle. To do this, first calculate percentages for each category. Then multiply each percentage by 360, the number of degrees in a circle, to get the number of degrees for the arc (part of circle) that represents each category's pie slice. For example, for the better prices category, multiply 37% by 360 degrees to get 133.2 degrees. Mark the endpoints of this arc on the circle using the protractor, and draw lines from the endpoints to the center of the circle. (If you draw your circle using a compass, the center of the circle can be easily identified.)

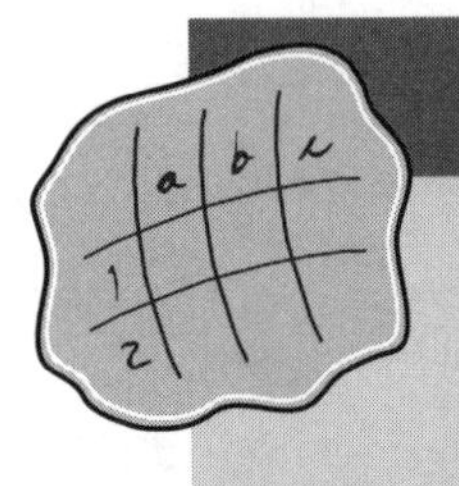

spreadsheet solution

Bar and Pie Charts

Chapter 2 Bar and **Chapter 2 Pie** present a bar and pie chart, respectively. Experiment with each chart by entering your own data in column B.

Best Practices

Sort your summary table data by the values in the second column before you create a chart. This will create a chart that fosters comparisons. For a bar chart, arrange values from smallest to largest value if you want the longest bar to appear at the top of the chart; otherwise, sort the values from largest to smallest.

Reformat charts created by software to eliminate unwanted gridlines and legends or to change the text font and size of titles and axis labels.

How-Tos

Tip CT1 (see Appendix D) explains how to sort data in a summary table.

Tip CT2 lists common chart-reformatting commands.

Tip CT3 lists the general steps for creating charts.

The Pareto Chart

CONCEPT A special type of bar chart that presents the counts, amounts, or percentages of each category in descending order left to right, and also contains a superimposed plotted line that represents a running cumulative percentage.

EXAMPLE

Causes of Incomplete ATM Transactions

Cause	Frequency	Percentage
ATM malfunctions	32	4.42%
ATM out of cash	28	3.87%
Invalid amount requested	23	3.18%
Lack of funds in account	19	2.62%
Card unreadable	234	32.32%
Warped card jammed	365	50.41%
Wrong keystroke	23	3.18%
Total	724	100.00%

Source: Data extracted from A. Bhalla, "Don't Misuse the Pareto Principle," *Six Sigma Forum Magazine*, May 2009, pp. 15–18.

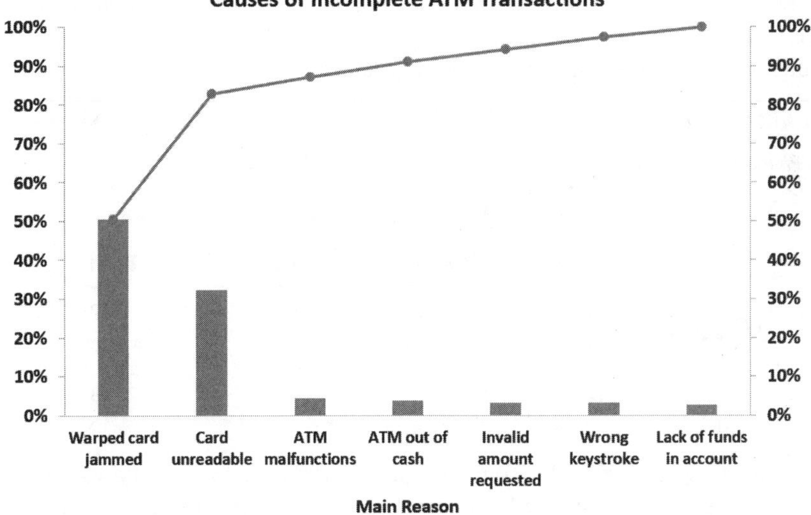

This Pareto chart uses the data of the table that immediately precedes it to highlight the causes of incomplete ATM transactions.

INTERPRETATION When you have many categories, a Pareto chart enables you to focus on the most important categories by visually separating the *vital few* from the *trivial many* categories. For the incomplete ATM transactions data, the Pareto chart shows that two categories, warped card jammed and card unreadable, account for more than 80% of all defects, and that those two categories combined with the ATM malfunctions and ATM out of cash categories account for more than 90% of all defects.

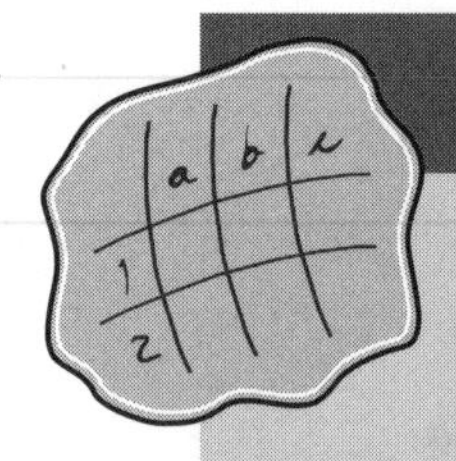

spreadsheet solution

Pareto Charts

Chapter 2 Pareto contains an example of a Pareto chart. Experiment with this chart by typing your own set of values—in descending order—in column B, rows 2 through 11. (Do not alter the entries in row 12 or columns C and D.)

How-To: Tip CT4 (see Appendix D) summarizes how to create a Pareto chart.

Two-Way Cross-Classification Table

CONCEPT A multicolumn table that presents the count or percentage of responses for two categorical variables. In a two-way table, the categories of one of the variables form the rows of the table, while the categories of the second variable form the columns. The "outside" of the table contains a special row and a special column that contain the totals. Cross-classification tables are also known as cross-tabulation tables.

EXAMPLES

Downloads Cross-Classified by Type of Call-to-Action Button

		Call-to-Action Button		
		Original	**New**	**Total**
Download	**Yes**	351	451	802
	No	3,291	3,105	6,396
	Total	3,642	3,556	7,198

This two-way cross-classification table summarizes the results of a webpage design study that investigated whether a new call to action button would increase the number of downloads. Tables showing row percentages, column percentages, and overall total percentages follow.

Row Percentages Table

		Call-to-Action Button		
		Original	**New**	**Total**
Download	**Yes**	43.77	56.23	100.00
	No	51.45	48.55	100.00
	Total	50.60	49.40	100.00

Column Percentages Table

		Call-to-Action Button		
		Original	New	Total
Download	Yes	9.64	12.68	11.14
	No	90.36	87.32	88.86
	Total	100.00	100.00	100.00

Overall Total Percentages Table

		Call-to-Action Button		
		Original	New	Total
Download	Yes	4.88	6.26	11.14
	No	45.72	43.14	88.86
	Total	50.60	49.40	100.00

INTERPRETATION The simplest two-way table contains a row variable that has two categories and a column variable that has two categories. This creates a table that has two rows and two columns in its inner part (see the illustration below). Each inner cell represents the count or percentage of a pairing, or cross-classifying, of categories from each variable.

	First Column Category	Second Column Category	Total
First Row Category	Count or percentage for first row and first column categories	Count or percentage for first row and second column categories	Total for first row category
Second Row Category	Count or percentage for second row and first column categories	Count or percentage for second row and second column categories	Total for second row category
Total	Total for first column category	Total for second column category	Overall total

Two-way tables can reveal the combination of values that occur most often in data. In this example, the tables reveal that the new call to action button is more likely to have downloads than the original call to action button. Because the number of visitors to each webpage was unequal in this example, you can see this pattern best in the Column Percentages table. That table shows that the new button is more likely to increase downloads than the original one.

PivotTables create worksheet summary tables from sample data and are a good way of creating a two-way table from sample data. Tip ADV1 in Appendix E discusses how to create these tables.

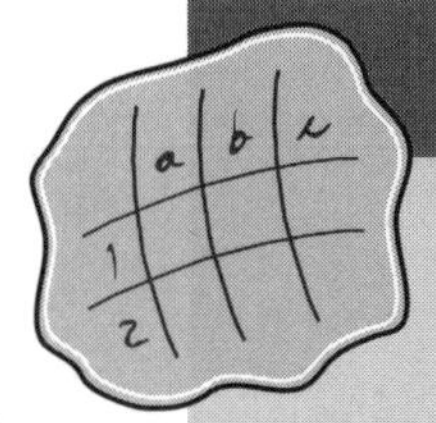

spreadsheet solution

Two-Way Tables

Chapter 2 Two-Way contains the counts of the download and call-to-action button variables as a simple two-way table.

Chapter 2 Two-Way PivotTable presents the counts of the download and call-to-action button variables summarized in a two-way table that is an Excel PivotTable.

How-To: Tip ADV1 in Appendix E summarizes how to create a PivotTable that is a two-way table.

2.2 Presenting Numerical Variables

You present numerical variables by first establishing groups that represent separate ranges of values and then placing each value into the proper group. Then you create tables that summarize the groups by frequency (count) or percentage and use the table as the basis for creating charts such as a histogram, which this chapter explains.

The Frequency and Percentage Distribution

CONCEPT A table of grouped numerical data that contains the names of each group in the first column, the counts (frequencies) of each group in the second column, and the percentages of each group in the third column. This table can also appear as a two-column table that shows either the frequencies or the percentages.

EXAMPLE The following Fan Cost Index data shows the cost (in $) for four tickets, two beers, four soft drinks, four hot dogs, two game programs, two caps, and one parking space at each of the 30 National Basketball Association arenas during a recent season.

NBACost

Arena	Fan Cost ($)	Arena	Fan Cost ($)
Atlanta	240.04	Miami	472.20
Boston	434.96	Milwaukee	309.30
Brooklyn	382.00	Minnesota	273.98
Charlotte	203.06	New Orleans	208.48
Chicago	456.60	New York	659.92
Cleveland	271.74	Oklahoma City	295.40
Dallas	321.18	Orlando	263.10
Denver	319.10	Philadelphia	266.40
Detroit	262.40	Phoenix	344.92
Golden State	324.08	Portland	308.18
Houston	336.05	Sacramento	268.28
Indiana	227.36	San Antonio	338.00
L.A. Clippers	395.20	Toronto	321.63
L.A. Lakers	542.00	Utah	280.98
Memphis	212.16	Washington	249.22

Source: Data extracted from FanCostExperience.com, **bit.ly/1nnu9rf**.

Remember that the file icon identifies a file that you can download for free from the website for this book (www.ftpress.com/evenyoucanlearnstatistics3e). See Appendix F for more information about downloading files.

The frequency and percentage distribution for the NBA Fan Cost Index is as follows:

Fan Cost ($)	Frequency	Percentage
200 to under 250	6	20.00%
250 to under 300	8	26.67%
300 to under 350	9	30.00%
350 to under 400	2	6.67%
400 to under 450	1	3.33%
450 to under 500	2	6.67%
500 to under 550	1	3.33%
550 to under 600	0	0.00%
600 to under 650	0	0.00%
650 to under 700	1	3.33%
	30	100.00%

INTERPRETATION Frequency and percentage distributions enable you to quickly determine differences among the many groups of values. In this example, you can quickly see that most of the fan cost indexes are between \$200 and \$350, and that very few fan cost indexes are above \$500.

You need to be careful in forming groups for distributions because the ranges of the group affect how you perceive the data. For example, had the fan cost indexes been grouped into only two groups, below \$300 and \$300 and above, you would not be able to see any pattern in the data.

Histogram

CONCEPT A special bar chart for grouped numerical data in which the frequencies or percentages in each group of numerical data are represented as individual bars on the vertical *Y* axis and the variable is plotted on the horizontal *X* axis. In a histogram, in contrast to a bar chart of categorical data, no gaps exist between adjacent bars.

EXAMPLE The following histogram presents the fan cost index data of the preceding example. The values below the bars (225, 275, 325, 375, 425, 475, 525, 575, 625, 675) are **midpoints**, the approximate middle value for each group of data. As with the frequency and percentage distributions, you can quickly see that very few fan cost indexes are above \$500.

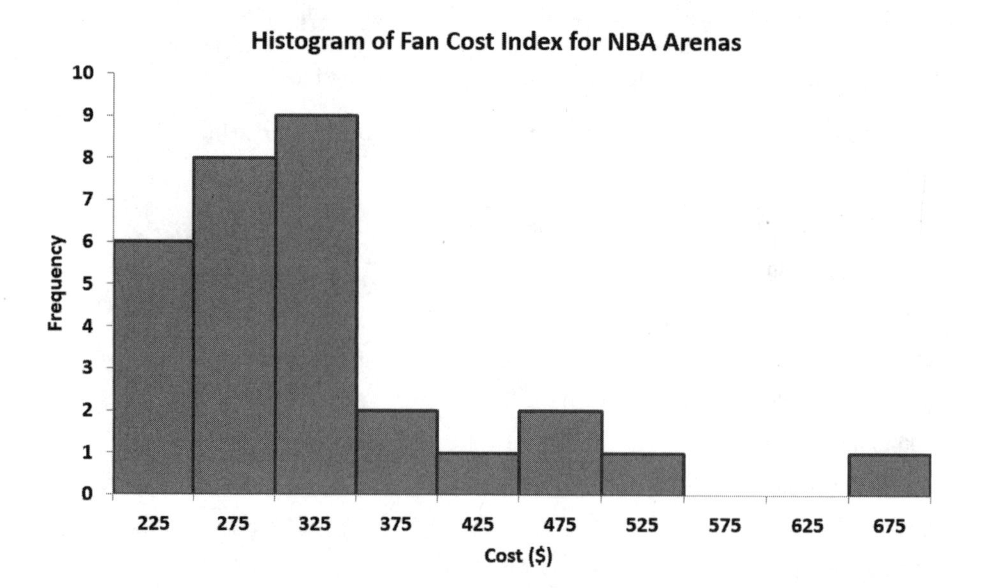

INTERPRETATION Histograms reveal the overall shape of the frequencies in the groups. Histograms are considered symmetric if each side of the chart is an approximate mirror image of the other side. (The histogram of this example has more values in the lower portion than in the upper portion, so it is considered to be skewed or non-symmetric.)

spreadsheet solution

Frequency Distributions and Histograms

Chapter 2 Histogram contains a frequency distribution and histogram for the Fan Cost Index (NBACost) data. Experiment with this chart by entering different values in column B, rows 3 through 12.

How-Tos: Tips ADV2, in Appendix E, and CT5, in Appendix D, discuss how you can create frequency distributions and histograms.

The Time-Series Plot

CONCEPT A chart in which each point represents the value of a numerical variable at a specific time. By convention, the *X* axis (the horizontal axis) always represents units of time, and the *Y* axis (the vertical axis) always represents units of the variable.

EXAMPLE The following data shows the annual revenues from movies from 1995 to 2013:

Movie Revenues

Year	Revenue ($ billions)	Year	Revenue ($ billions)
1995	5.29	2005	8.93
1996	5.59	2006	9.25
1997	6.51	2007	9.63
1998	6.78	2008	9.95
1999	7.30	2009	10.65
2000	7.48	2010	10.54
2001	8.13	2011	10.19
2002	9.19	2012	10.83
2003	9.35	2013	9.77
2004	9.11		

Source: Data extracted from www.the-numbers.com/market, February 12, 2014.

The time-series plot of these data follows.

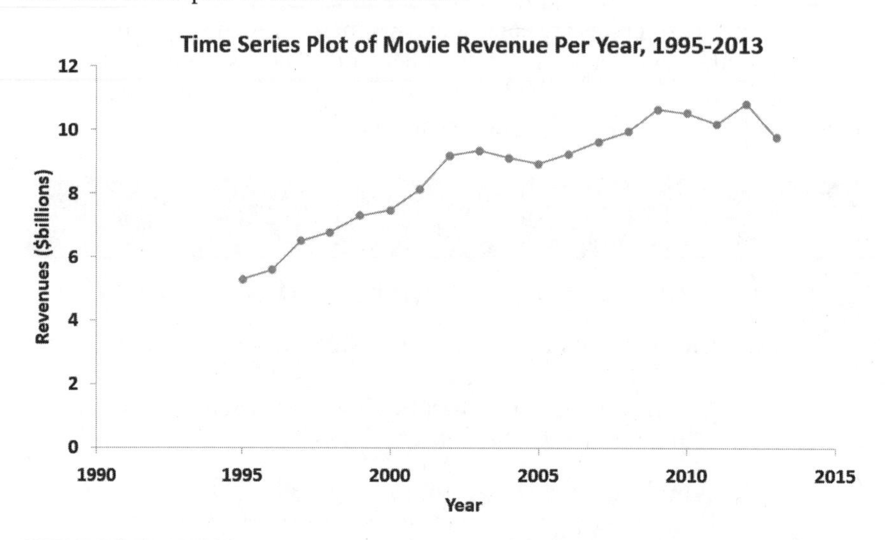

INTERPRETATION Time-series plots can reveal patterns over time, patterns that you might not see when looking at a long list of numerical values. In this example, the plot reveals that there was a steady increase in the revenue of movies between 1995 and 2003, a leveling off from 2003 and 2006, followed by a further increase from 2007 to 2009, followed by another leveling off from 2010 to 2012, and then a decline in 2013 back to the level below the revenue for 2008. During that time, the revenue increased from under $6 billion in 1995 to more than $10 billion in 2009 to 2012.

The Scatter Plot

CONCEPT A chart that plots the values of two numerical variables for each observation. In a scatter plot, the *X* axis (the horizontal axis) always represents units of one variable, and the *Y* axis (the vertical axis) always represents units of the second variable.

EXAMPLE The following data tabulates the labor hours used and the cubic feet of material moved for 36 moving jobs:

Moving

Job	Labor Hours	Cu. Feet	Job	Labor Hours	Cu. Feet
M-1	24.00	545	M-19	25.00	557
M-2	13.50	400	M-20	45.00	1,028
M-3	26.25	562	M-21	29.00	793
M-4	25.00	540	M-22	21.00	523
M-5	9.00	220	M-23	22.00	564
M-6	20.00	344	M-24	16.50	312
M-7	22.00	569	M-25	37.00	757

M-8	11.25	340	M-26	32.00	600
M-9	50.00	900	M-27	34.00	796
M-10	12.00	285	M-28	25.00	577
M-11	38.75	865	M-29	31.00	500
M-12	40.00	831	M-30	24.00	695
M-13	19.50	344	M-31	40.00	1,054
M-14	18.00	360	M-32	27.00	486
M-15	28.00	750	M-33	18.00	442
M-16	27.00	650	M-34	62.50	1,249
M-17	21.00	415	M-35	53.75	995
M-18	15.00	275	M-36	79.50	1,397

The scatter plot of these data follows.

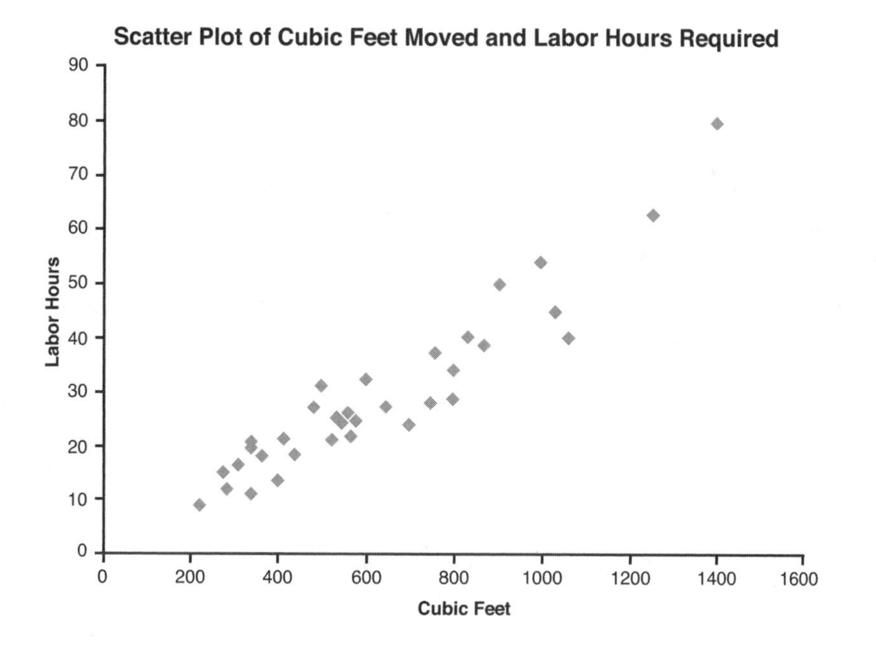

INTERPRETATION Scatter plots help reveal patterns in the relationship between two numerical variables. The scatter plot for these data reveals a strong positive linear (straight line) relationship between the number of cubic feet moved and the number of labor hours required. Observing this relationship, you can conclude that the number of cubic feet being moved in a specific job is a useful predictor of the number of labor hours that are needed. Using one numerical variable to predict the value of another is more fully discussed in Chapter 10.

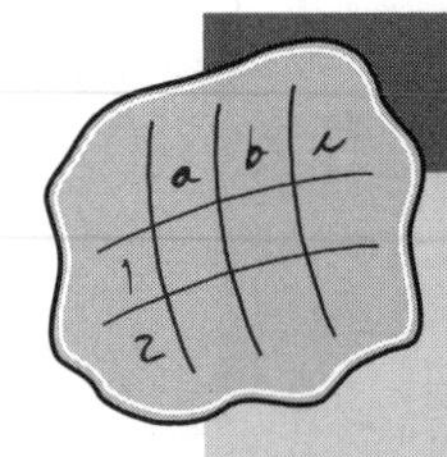

spreadsheet solution

Time-Series and Scatter Plots

Chapter 2 Time-Series contains the time-series plot for the annual movie revenues data. Experiment with this plot by entering different values in column B, rows 2 through 20.

How-To: Tip CT6 (in Appendix D) discusses how you can create time-series plots.

Chapter 2 Scatter Plot contains the scatter plot for the moving jobs data. Experiment with this scatter plot by entering different data values in columns B and C, rows 2 through 37.

How-To: Tip CT7 discusses how you can create scatter plots.

2.3 "Bad" Charts

So-called "good" charts, such as the charts presented so far in this chapter, help visualize data in ways that aid understanding. However, in the modern world, you can easily find examples of "bad" charts that obscure or confuse the data. Such charts fail to apply properly the techniques discussed in this chapter or include elements or practices know to impede understanding.

CONCEPT A "bad" chart fails to clearly present data in a useful and undistorted manner.

INTERPRETATION Using pictorial symbols obscures the data and can create a false impression in the mind of the reader, especially if the pictorial symbols are representations of three-dimensional objects. In Example 1, the wine glasses fail to reflect that the 1992 data (2.25 million gallons) is a bit more than twice the 1.04 million gallons for 1989. In addition, the spaces between the wine glasses falsely suggest equal-sized time periods and obscures the trend in wine exports. (Hint: Plot this data as a time-series chart to discover the actual trend.)

EXAMPLE 1: Australian Wine Exports to the United States.

We're drinking more...
Australian wine exports to the U.S.
in millions of gallons

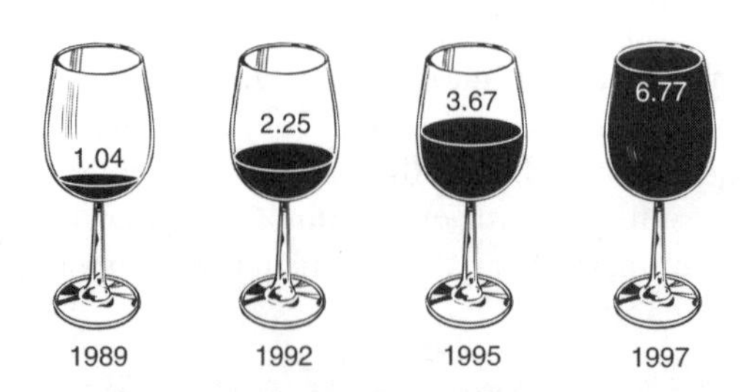

Example 2 combines the inaccuracy of using a picture (grape vine) with the error of having unlabeled and improperly scaled axes. A missing *X* axis prevents the reader from immediately seeing that the 1997–1998 value is misplaced. By the scale of the graph, that data point should be closer to the rest of the data. A missing *Y* axis prevents the reader from getting a better sense of the rate of change in land planted through the years. Other problems also exist. Can you spot at least one more? (Hint: Compare the 1949–1950 data to the 1969–1970 data.)

EXAMPLE 2: Amount of Land Planted with Grapes for the Wine Industry.

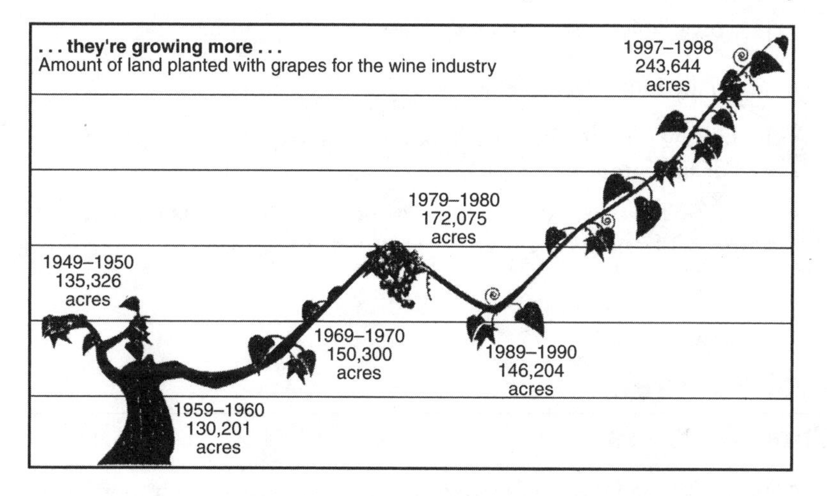

When producing your own charts, use these guidelines:

- Always choose the simplest chart that can present your data.
- Always supply a title.
- Always label every axis.
- Avoid unnecessary decorations or illustrations around the borders or in the background.
- Avoid the use of fancy pictorial symbols to represent data values.
- Avoid 3D versions of bar and pie charts.
- If the chart contains axes, always include a scale for each axis.
- When charting non-negative values, the scale on the vertical axis should begin at zero.

One-Minute Summary

To choose an appropriate table or chart type, begin by determining whether your data are categorical or numerical.

If your data are categorical:

- Determine whether you are presenting one or two variables.
- If one variable, use a summary table and/or bar chart, pie chart, or Pareto chart.
- If two variables, use a two-way cross-classification table.

If your data are numerical:

- If charting one variable, use a frequency and percentage distribution and/or histogram.
- If charting two variables, if the time order of the data is important, use a time-series plot; otherwise, use a scatter plot.

Test Yourself

Short Answers

1. Which of the following graphical presentations is not appropriate for categorical data?
   (a) Pareto chart
   (b) scatter plot
   (c) bar chart
   (d) pie chart
2. Which of the following graphical presentations is not appropriate for numerical data?
   (a) histogram
   (b) pie chart
   (c) time-series plot
   (d) scatter plot
3. A type of histogram in which the categories are plotted in the descending rank order of the magnitude of their frequencies is called a:
   (a) bar chart
   (b) pie chart
   (c) scatter plot
   (d) Pareto chart

4. Which of the following would best show that the total of all the categories sums to 100%?
   (a) pie chart
   (b) histogram
   (c) scatter plot
   (d) time-series plot

5. The basic principle behind the ________ is the capability to separate the vital few categories from the trivial many categories.
   (a) scatter plot
   (b) bar chart
   (c) Pareto chart
   (d) pie chart

6. When studying the simultaneous responses to two categorical variables, you should construct a:
   (a) histogram
   (b) pie chart
   (c) scatter plot
   (d) cross-classification table

7. In a cross-classification table, the number of rows and columns:
   (a) must always be the same
   (b) must always be 2
   (c) must add to 100%
   (d) None of the above.

Answer True or False:

8. Histograms are used for numerical data, whereas bar charts are suitable for categorical data.

9. A website monitors customer complaints and organizes these complaints into six distinct categories. Over the past year, the company has received 534 complaints. One possible graphical method for representing these data is a Pareto chart.

10. A website monitors customer complaints and organizes these complaints into six distinct categories. Over the past year, the company has received 534 complaints. One possible graphical method for representing these data is a scatter plot.

11. A social media website collected information on the age of its customers. The youngest customer was 5 and the oldest was 96. To study the distribution of the age of its customers, the company should use a pie chart.

12. A social media website collected information on the age of its customers. The youngest customer was 5, and the oldest was 96. To study the distribution of the age of its customers, the company can use a histogram.
13. A website wants to collect information on the daily number of visitors. To study the daily number of visitors, it can use a pie chart.
14. A website wants to collect information on the daily number of visitors. To study the daily number of visitors, it can use a time-series plot.
15. A professor wants to study the relationship between the number of hours a student studied for an exam and the exam score achieved. The professor can use a time-series plot.
16. A professor wants to study the relationship between the number of hours a student studied for an exam and the exam score achieved. The professor can use a bar chart.
17. A professor wants to study the relationship between the number of hours a student studied for an exam and the exam score achieved. The professor can use a scatter plot.
18. If you wanted to compare the percentage of items that are in a particular category as compared to other categories, you should use a pie chart, not a bar chart.

Fill in the blank:

19. To evaluate two categorical variables at the same time, a _______ should be developed.
20. A _______ is a vertical bar chart in which the rectangular bars are constructed at the boundaries of each class interval.
21. A _______ chart should be used when you are primarily concerned with the percentage of the total that is in each category.
22. A _______ chart should be used when you are primarily concerned with comparing the percentages in different categories.
23. A _______ should be used when you are studying a pattern between two numerical variables.
24. A _______ should be used to study the distribution of a numerical variable.
25. You have measured your pulse rate daily for 30 days. A _______ plot should be used to study the pulse rate for the 30 days.
26. You have collected data from your friends concerning their favorite soft drink. You should use a _______ chart to study the favorite soft drink of your friends.
27. You have collected data from your friends concerning the time it takes to get ready to leave their house in the morning. You should use a _______ to study this variable.

Answers to Test Yourself Short Answers

1. b
2. b
3. d
4. a
5. c
6. d
7. d
8. True
9. True
10. False
11. False
12. True
13. False
14. True
15. False
16. False
17. True
18. False
19. cross-classification table
20. histogram
21. pie chart
22. bar chart
23. scatter plot
24. histogram
25. time-series plot
26. bar chart, pie chart, or Pareto chart
27. histogram

Problems

1. A survey of 503 workers asked when they expected a response to email ("Answering email," *USA Today*, June 6, 2013, p.1A). The results were as follows:

Time for Expected Email Response	Percentage (%)
30 minutes or less	31
An hour	22
Few hours	19
Working day	23
Week or more	5

   (a) Construct a bar chart and a pie chart.

   (b) Which graphical method do you think is best to portray these data?

   (c) What conclusions can you reach concerning the time within which people expect an email response?

2. Medication errors are a serious problem in hospitals. The following data represent the root causes of pharmacy errors at a hospital during a recent time period:

Reason for Failure	Frequency
Additional instructions	16
Dose	23
Drug	14
Duplicate order entry	22
Frequency	47
Omission	21
Order not discontinued when received	12
Order not received	52
Patient	5
Route	4
Other	8

(a) Construct a Pareto chart.

(b) Discuss the "vital few" and "trivial many" reasons for the root causes of pharmacy errors.

DomesticBeer

3. The file **DomesticBeer** contains the percentage alcohol, number of calories per 12 ounces, and number of carbohydrates (in grams) per 12 ounces for 152 of the best-selling domestic beers in the United States. (Data extracted from **www.beer100.com/beercalories.htm**, March 20, 2013.)

   (a) Construct a frequency distribution and a percentage distribution for percentage alcohol, number of calories per 12 ounces, and number of carbohydrates per 12 ounces (in grams).

   (b) Construct a histogram for percentage alcohol, number of calories per 12 ounces, and number of carbohydrates per 12 ounces (in grams).

   (c) Construct three scatter plots: percentage alcohol versus calories, percentage alcohol versus carbohydrates, and calories versus carbohydrates.

   (d) What conclusions can you reach about the percentage alcohol, number of calories per 12 ounces, and number of carbohydrates per 12 ounces (in grams)?

PropertyTaxes

4. The **PropertyTaxes** file contains the property taxes per capita for the 50 states and the District of Columbia.

   (a) Construct a histogram.

   (b) What conclusions can you reach concerning the property taxes per capita?

5. The following table shows the per capita consumption of bottled water in the United States from 2009 to 2013:

Bottled Water

Year	Per Capita Consumption of Bottled Water
2009	27.6
2010	28.3
2011	29.2
2012	30.8
2013	32.0

Source: Data extracted from "American Thirst for Bottled Water," *USA Today*, January 22, 2014, p. 1A.

(a) Construct a time-series plot for the per capita consumption of bottled water in the United States from 2009 to 2013.

(b) What pattern, if any, is present in the data?

(c) If you had to make a prediction of the per capita consumption of bottled water in the United States in 2014, what would you predict?

6. The following data shows the calories and sugar (in grams) of seven breakfast cereals.

Cereals

Cereal	Calories	Sugar
Kellogg's All Bran	80	6
Kellogg's Corn Flakes	100	2
Wheaties	100	4
Nature's Path Organic Multigrain Flakes	110	4
Kellogg's Rice Crispies	130	4
Post Shredded Wheat Vanilla Almond	190	11
Kellogg's Mini Wheats	200	10

(a) Construct a scatter plot.

(b) What conclusion can you reach about the relationship between calories and sugar in these cereals?

Answers to Test Yourself Problems

1. (b) If you are more interested in determining which category of email response occurs most often, then the bar chart is preferred. If you are more interested in seeing the distribution of the entire set of categories, the pie chart is preferred.

(c) More than half the workers expect a response within an hour. Very few workers expect a response in a week or more.

2. (b) The most important categories of medication errors are orders not received and frequency followed by dose, duplicate order entry, and omission.
3. (c) The alcohol percentage is concentrated between 4% and 6%, with more between 4% and 5%. The calories are concentrated between 140 and 160. The carbohydrates are concentrated between 12 and 15. You can see outliers in the percentage of alcohol in both tails. The outlier in the lower tail is due to the nonalcoholic beer O'Doul's. The outlier in the upper tail is around 11.5%. A few beers have calorie content as high as around 327.5 and carbohydrates as high as 31.5. A strong positive relationship exists between percentage of alcohol and calories and between calories and carbohydrates, and there is a moderately positive relationship between percentage alcohol and carbohydrates.
4. (b) Property taxes seem concentrated between $1,000 and $1,500 and also between $500 and $1,000 per capita. More states have property taxes per capita below $1,500 than above $1,500.
5. (b) You can see an upward trend in the amount of water consumed between 2009 and 2013.

   (c) With extrapolation, you would predict the per capita consumption of bottled water to be about 33 gallons in 2014 because consumption is increasing by about 1 gallon per year.
6. (b) A strong positive relationship exists between the number of calories and the amount of sugar. Cereals that have more calories have higher amounts of sugar.

References

1. Beninger, J. M., and D. L. Robyn. 1978. "Quantitative Graphics in Statistics." *The American Statistician*, 32: 1–11.
2. Berenson, M. L., D. M. Levine, and K. A. Szabat. *Basic Business Statistics: Concepts and Applications, Thirteenth Edition*. Upper Saddle River, NJ: Pearson Education, 2015.
3. Microsoft Excel 2013. Redmond, WA: Microsoft Corporation, 2012.
4. Tufte, E. R. *Beautiful Evidence*. Cheshire, CT: Graphics Press, 2006.
5. Tufte, E. R. *Envisioning Information*. Cheshire, CT: Graphics Press, 1990.
6. Tufte, E. R. *The Visual Display of Quantitative Information*, 2nd ed. Cheshire, CT: Graphics Press, 2002.
7. Tufte, E. R. *Visual Explanations*. Cheshire, CT: Graphics Press, 1997.

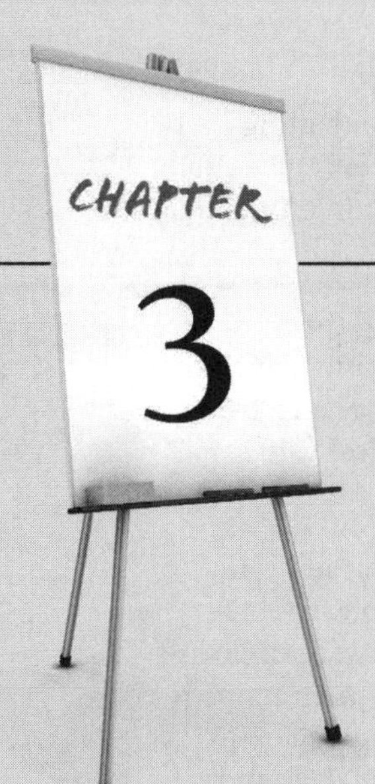

Descriptive Statistics

Besides the tables and charts discussed in Chapter 2, you can also summarize and describe numerical variables using descriptive measures that identify the properties of central tendency, variation, and shape.

3.1 Measures of Central Tendency

The data values for most numerical variables tend to group around a specific value. **Measures of central tendency** help describe to what extent this pattern holds for a specific numerical variable. This section discusses three commonly-used measures: the arithmetic mean, also known as the mean or average, the median, and the mode. You calculate these measures as either sample statistics or population parameters.

The Mean

CONCEPT A number equal to the sum of the data values, divided by the number of data values that were summed.

EXAMPLES Many common sports statistics such as baseball batting averages and basketball points per game, mean score on a college entrance exam, mean age of the members of a social media website, mean waiting times at a bank.

INTERPRETATION The mean represents one way of finding the most typical value in a set of data values. As the only measure of central tendency that uses all the data values in a sample or population, the mean has one great weakness: Individual extreme values can distort the most typical value, as WORKED-OUT PROBLEM 2 illustrates.

WORKED-OUT PROBLEM 1 Although many people sometimes find themselves running late as they get ready to go to work, few measure the actual time it takes to get ready in the morning. Suppose you want to determine the typical time that elapses between your alarm clock's programmed wake-up time and the time you leave your home for work. You decide to measure actual times (in minutes) for ten consecutive working days and record the following times:

Times

Day	1	2	3	4	5	6	7	8	9	10
Time	39	29	43	52	39	44	40	31	44	35

To compute the mean time, first compute the sum of all the data values: 39 + 29 + 43 + 52 + 39 + 44 + 40 + 31 + 44 + 35, which is 396. Then, take this sum of 396 and divide by 10, the number of data values. The result, 39.6 minutes, is the mean time to get ready.

WORKED-OUT PROBLEM 2 Consider the same problem but imagine that on day 4, an exceptional occurrence such as an unexpected phone call caused you to take 102 (and not 52) minutes to get ready in the morning. That would make the sum of all times, 446 minutes, and the mean (446 divided by 10), 44.6 minutes.

This illustrates how one extreme value can dramatically change the mean. Instead of being a number at or near the middle of the ten get-ready times, the new mean of 44.6 minutes is greater than 9 of the 10 get-ready times. In this case, the mean fails as a measure of a typical value or "central tendency."

The Median

CONCEPT The middle value when a set of the data values have been ordered from lowest to highest value. When the number of data values is even, no natural middle value exists and you calculate the mean of the two middle values to determine the median as the Interpretation on page 40 explains.

equation blackboard (optional)

interested in math?

The WORKED-OUT PROBLEMS calculate the mean of a sample of get-ready times. You need three symbols to write the equation for calculating the mean:

- An uppercase italic *X* with a horizontal line above it, $\bar{X}$, pronounced as "*X* bar," that represents the number that is the mean of a sample.
- A subscripted uppercase italic *X* (for example, X_1) that represents one of the data values being summed. Because the problem contains ten data values, there are ten *X* values, the first one labeled X_1, the last one labeled X_{10}.
- A lowercase italic *n* that represents the number of data values that were summed in this sample, a concept also known as the **sample size**. You pronounce *n* as "sample size" to avoid confusion with the symbol *N* that represents (and is pronounced as) the population size.

Using these symbols creates the following equation:

$$\bar{X} = \frac{X_1 + X_2 + X_3 + X_4 + X_5 + X_6 + X_7 + X_8 + X_9 + X_{10}}{n}$$

By using an ellipsis (...), you can abbreviate the equation as:

$$\bar{X} = \frac{X_1 + X_2 + \cdots + X_{10}}{n}$$

Using the insight that the value of the last subscript will always be equal to the value of *n*, you can generalize the formula as:

$$\bar{X} = \frac{X_1 + X_2 + \cdots + X_n}{n}$$

By using the uppercase Greek letter sigma, Σ, a standard symbol that is used in mathematics to represent the summing of values, you can further simplify the formula as

$$\bar{X} = \frac{\Sigma X}{n}$$

or more explicitly as

$$\bar{X} = \frac{\sum_{i=1}^{n} X_i}{n}$$

in which *i* represents a placeholder for a subscript and the $i = 1$ and *n* below and above the sigma represent the numerical range of the subscripts to be used in the calculation.

EXAMPLES Economic statistics such as median household income for a region; marketing statistics such as the median age for purchasers of a consumer product; in education, the established middle point for many standardized tests.

INTERPRETATION The median splits the set of ordered data values into two parts that have an equal number of values. The median is a good alternative to the mean when extreme data values occur because, unlike the mean, extreme values do not affect the median.

When you summarize an odd number of values, you calculate the median as the middle value of the ordered lowest to highest list of all data values. For example, if you had five ordered values, the median would be the third ordered value.

When you summarize an even number of data values, you calculate the median by taking the mean of the two values closest to the middle of the ordered lowest-to-highest list of all data values. For example, if you had six ordered values, you would calculate the mean of the third and fourth values. If you had ten ordered values, you would calculate the mean of the fifth and sixth values. (Some people use the term *ranks* to refer to ordered values, so the statement "If you had ten ranked values, you would calculate the mean of the fifth and sixth ranks" is equivalent to the previous sentence.)

important point

When a numerical variable has a very large number of data values, you cannot easily identify by visual inspection the middle value (or the middle two values, if the number of data values is even). When faced with a very large number of data values, add 1 to the number of data values and divide that sum by 2 to identify the position, or *rank*, of the median value in the ordered list of data values for the variable. For example, if you had 127 ordered data values, you would divide 128 by 2 to get 64, to identify the median as the 64th ranked value. If you had 70 ordered values, you would divide 71 by 2 to get 35.5, to determine that the median is the mean of the 35th and 36th ranked values.

WORKED-OUT PROBLEM 3 You need to determine the median age of a group of employees whose individual ages are 47, 23, 34, 22, and 27. You calculate the median by first ranking the ages from lowest to highest: 22, 23, 27, 34, and 47. Because you have five values, the middle is the third ranked value, 27, making the median 27. This means that half the workers are 27 years old or younger and half the workers are 27 years old or older.

WORKED-OUT PROBLEM 4 You need to determine the median for the original set of ten get-ready times from WORKED-OUT PROBLEM 1 on page 38. Ordering these values from lowest to highest, you have

Time	29	31	35	39	39	40	43	44	44	52
Ordered Position	1st	2nd	3rd	4th	5th	6th	7th	8th	9th	10th

Because an even number of data values exists (ten), you calculate the mean of the two values closest to the middle—that is, the fifth and sixth ranked values, 39 and 40. The mean of 39 and 40 is 39.5, making the median 39.5 minutes for the set of ten times to get ready.

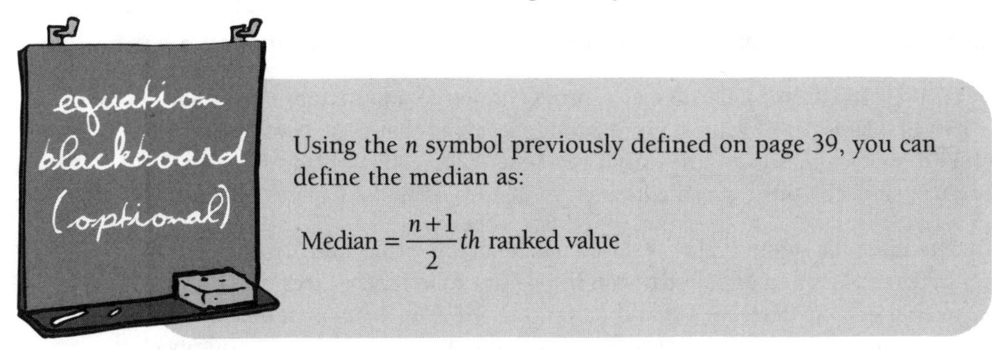

Using the n symbol previously defined on page 39, you can define the median as:

$$\text{Median} = \frac{n+1}{2}\textit{th} \text{ ranked value}$$

The Mode

CONCEPT The value (or values) in a set of data values that appears most frequently.

EXAMPLES The most common score on an exam, the most common number of items purchased in one transaction at a store, the commuting time that occurs most often.

INTERPRETATION Some sets of data values have no mode—all the unique values appear the same number of times. Other sets of data values can have more than one mode, such as the get-ready times on page 38. This set contains two modes, 39 minutes and 44 minutes, because each of these values appears twice and all other values appear once.

Like the median, extreme values do not affect the mode. However, unlike the median, the mode can vary much more from sample to sample than the median or the mean.

3.2 Measures of Position

Measures of position describe the relative position of a data value of a numerical variable to the other values of the variable. One commonly used measure of position are the quartiles. (As first noted on page 40, some use the term *ranks* to refer to ordered positions. As using ranks is the preferred usage when discussing quartiles, this section uses ranks when referring to ordered positions.)

Quartiles

CONCEPT The three values that split a set of ranked data values into four equal parts, or quartiles. The **first quartile,** Q_1, is the value such that 25.0%

of the ranked data values are smaller and 75.0% are larger. The **second quartile, Q_2**, splits the ranked values into two equal parts. (The second quartile is another name for the median, defined on page 38.) The **third quartile, Q_3**, is the value such that 75.0% of the ranked values are smaller and 25.0% are larger.

EXAMPLE Standardized tests that report results in terms of quartiles.

INTERPRETATION Quartiles help bring context to a particular value that is part of a large set of values. For example, learning that you scored 580 (out of 800) on a standardized test would not be as informative as learning that you scored in the third quartile, that is, in the top 25% of all scores.

To determine the ranked value that defines the first quartile, add 1 to the number of data values and divide that sum by 4. For example, for 11 values, divide 12 by 4 to get 3 to determine that the third ranked value is the first quartile. To determine the ranked value that defines the third quartile, add 1 to the number of data values, divide that sum by 4, and multiply the quotient by 3. For the example of 11 values, the ninth ranked value is the third quartile (12 divided by 4 is 3 and 3 times 3 is 9). To determine the ranked value that defines the second quartile, use the page 40 instructions for calculating the median.

When the result of a quartile rank calculation is not an integer:

1. Select the ranked value whose rank is immediately below the calculated rank and select the ranked value whose rank is immediately above the calculated rank. For example, if the result of a quartile rank calculation is 3.75, select the third and fourth ranked values.
2. If the two ranked values selected are the same number, then the quartile is that number. If the two values are different numbers, continue with steps 3 through 5.
3. Multiply the larger ranked value by the decimal fraction of the calculated rank. (The decimal fraction will be either 0.25, 0.50, or 0.75.)
4. Multiply the smaller ranked value by 1 minus the decimal fraction of the calculated rank.
5. Add the two products to determine the quartile value.

For example, if you had ten values, the calculated rank for the first quartile would be 2.75 (10 + 1 is 11, 11/4 is 2.75). Because 2.75 is not an integer, you would select the second and third ranked values. If these two values were the same, then the first quartile would be the shared value. Otherwise, by steps 3 and 4, you multiply the third ranked value by 0.75 and multiply the second ranked value by 0.25 (1 – 0.75 is 0.25). Then, by step 5, you would add these products together to get the first quartile.

For measures such as standardized test scores, another statistic, the **percentile**, is often used in addition to the quartile. The percentile expresses the percentage of ranked values that are lower than the result being reported. By the definitions given earlier, the first quartile, Q_1, is the 25th percentile;

the second quartile, Q_2, is the 50th percentile; and third quartile, Q_3, is the 75th percentile. A score reported as being in the 99th percentile would be exceptional because that score is greater than 99% of all scores; that is, the score is in the top 1% of all scores.

WORKED-OUT PROBLEM 5 You are asked to determine the first quartile for the ranked get-ready times first shown on page 40 and shown here.

Time	29	31	35	39	39	40	43	44	44	52
Ranked Value	1st	2nd	3rd	4th	5th	6th	7th	8th	9th	10th

You first calculate the rank for the first quartile as 2.75 (10 + 1 is 11, 11/4 is 2.75). Because 2.75 is not an integer, you select the second and third ranked values. By steps 3 and 4 on page 42, you multiply the third ranked value, 35, by 0.75 to get 26.25 and multiply the second ranked value, 31, by 0.25 to get 7.75. By step 5, you calculate the first quartile as 34 (26.25 + 7.75 is 34). This means that 25% of the get-ready times are 34 minutes or less and that the other 75% are 34 minutes or more.

WORKED-OUT PROBLEM 6 You are asked to determine the third quartile for the ranked get-ready times. Calculate the rank for the third quartile as 8.25 (10 + 1 is 11, 11/4 is 2.75, 2.75 times 3 is 8.25). Because 8.25 is not an integer, you select the eighth and ninth ranked values. By step 2, because both of these values are 44, the third quartile is 44. (No multiplication is necessary.)

equation blackboard (optional)

interested in math?

Using the equation for the median developed earlier,

$$\text{Median} = \frac{n+1}{2} \text{ ranked value,}$$

you can express the **first quartile,** Q_1, as

$$Q_1 = \frac{n+1}{4} th \text{ ranked value}$$

and the **third quartile,** Q_3, as

$$Q_3 = \frac{3(n+1)}{4} th \text{ ranked value}$$

WORKED-OUT PROBLEM 7 You conduct a study that compares the cost for a restaurant meal in a major city to the cost of a similar meal in the suburbs outside the city. You collect meal cost per person data from a sample of 50 city restaurants and 50 suburban restaurants and arrange the 100 values in two ranked sets as follows:

Restaurants

City Restaurant Meal Cost Data

25 26 27 29 32 32 33 33 34 35 35 36 37 39 41 42 42 43 43 43 44 44 44 44 45 48 50 50 50 50 51 53 54 55 56 57 57 60 61 61 65 66 67 68 74 74 76 77 77 80

Suburban Restaurant Meal Cost Data

26 27 28 29 31 33 34 34 34 34 34 34 35 36 37 37 37 38 39 39 39 40 41 41 43 44 44 44 46 47 47 48 48 49 50 51 51 51 51 52 52 54 56 59 60 60 67 68 70 71

Because you have two sets of 50 ordered numbers, you decide to create a spreadsheet similar to the one discussed in the Spreadsheet Solutions, on page 45. You enter the city and suburban restaurant meal cost data in separate columns, using the row 1 cells for the column headings (City and Suburban). You enter names of various descriptive measures in another column and then enter formulas to calculate those measures in subsequent columns. You create a spreadsheet similar to the one in **Chapter 3 Worked-out Problem 7**, part of which is shown below.

Descriptive Statistics		
Measures of Central Tendency		
	City	Suburban
Arithmetic Mean	49.3	44.4
Median	46.5	43.5
Mode	44	34
only the first mode is reported		
Measures of Position		
	City	Suburban
First Quartile	36.75	34.75
Third Quartile	60.25	51.00

From the spreadsheet results, you note that:

- The mean cost of city meals, \$49.30, is higher than the mean cost of suburban meals, \$44.40.
- The median cost of a city meal, \$46.50, is higher than the median suburban cost, \$43.50.
- The first and third quartiles for city meals (\$36.75 and \$60.25) are higher than for suburban meals (\$34.75 and \$51.00).

From these results, you conclude that the cost of a restaurant meal per person is higher in the city than it is in the suburbs.

spreadsheet solution

Measures of Central Tendency and Position

Chapter 3 Descriptive contains a spreadsheet (shown below) that calculates measures of central tendency and position for the sample of get-ready times. Experiment with the spreadsheet by entering your own data in column B.

Best Practices

Enter the data for the variable being summarized in its own column.

Use the **AVERAGE** (for the mean), **MEDIAN**, and **MODE** functions in worksheet formulas to calculate measures of central tendency.

Use the **QUARTILE.EXC** function to calculate the first and third quartiles.

How-Tos

Tip FT1 in Appendix D explains how to enter the functions used in **Chapter 3 Descriptive**.

Tip ATT2 in Appendix E describes using the Analysis ToolPak as a second way to generate measures of central tendency.

	A	B	C	D	
1	Data		Descriptive Statistics		
2	29				
3	31		Measures of Central Tendency		
4	35		Arithmetic Mean	39.6	=AVERAGE(A:A)
5	39		Median	39.5	=MEDIAN(A:A)
6	39		Mode	39	=MODE(A:A)
7	40		*only the first mode is reported*		
8	43				
9	44		Measures of Position		
10	44		First Quartile	34.00	=QUARTILE.EXC(A:A,1)
11	52		Third Quartile	44.00	=QUARTILE.EXC(A:A,3)
12					
13			Measures of Variation		
14			Maximum	52	=MAX(A:A)
15			Minimum	29	=MIN(A:A)
16			Range	23	=D14 - D15
17			Variance	45.82	=VAR.S(A:A)
18			Standard Deviation	6.77	=STDEV.S(A:A)

3.3 Measures of Variation

Measures of **variation** show the amount of **dispersion**, or spread, in the data values of a numerical variable. Four frequently used measures of variation are the range, the variance, the standard deviation, and the Z score, all of which can be calculated as either sample statistics or population parameters.

The Range

CONCEPT The difference between the largest and smallest values in a set of data values.

EXAMPLES The daily high and low temperatures, the stock market 52-week high and low closing prices, the fastest and slowest times for timed sporting events.

INTERPRETATION The range is the number that represents the largest possible difference between any two values in a set of data values. The greater the range, the greater the variation in the data values.

WORKED-OUT PROBLEM 8 For the get-ready times data first presented on page 38, the range is 23 minutes (52–29). For the restaurant meal study, for the city meal cost data, the range is \$55, and the range for the suburban meal cost data is \$45. You can conclude that meal costs in the city show more variation than suburban meal costs.

For a set of data values, the range is equal to

Range = largest value – smallest value

The Variance and the Standard Deviation

CONCEPT Two measures that describe how the set of data values for a variable are distributed around the mean of the variable. The standard deviation is the positive square root of the variance.

EXAMPLE The variance among SAT scores for incoming freshmen at a college, the standard deviation of the time visitors spend on a website, the standard deviation of the annual return of a certain type of mutual funds.

INTERPRETATION For almost all sets of data values, most values lie within an interval of plus and minus one standard deviation above and below the mean. Therefore, determining the mean and the standard deviation usually helps you define the range in which the majority of the data values occur.

interested in math?

To calculate the variance, you take the difference between each data value and the mean, square this difference and then sum the squared differences. You then take this sum of squares (or *SS*) and divide it by 1 less than the number of data values, for sample data, or the number of data values, for population data. The result is the variance.

Because calculating the variance includes squaring the difference between each value and the mean, a step that always produces a non-negative number, the variance can never be negative. And as the positive square root of such a non-negative number, the standard deviation can never be negative, either.

WORKED-OUT PROBLEM 9 You want to calculate the variance and standard deviation for the get-ready times first presented on page 38. As first steps, you calculate the difference between each of the 10 individual times and the mean (39.6 minutes), square those differences, and sum the squares (shown below).

Day	Time	Difference: Time Minus the Mean	Square of the Difference
1	39	−0.6	0.36
2	29	−10.6	112.36
3	43	3.4	11.56
4	52	12.4	153.76
5	39	−0.6	0.36
6	44	4.4	19.36
7	40	0.4	0.16
8	31	−8.6	73.96
9	44	4.4	19.36
10	35	−4.6	21.16
		Sum of Squares:	412.40

Because this is a sample of get-ready times, the sum of squares, 412.40, is divided by one less than the number of data values, 9, to get 45.82, the sample variance. The square root of 45.82 (6.77, after rounding) is the sample standard deviation. You can then reasonably conclude that most get-ready times are between 32.83 (39.6 – 6.77) minutes and 46.37 (39.6 + 6.77) minutes, a statement that you can easily confirm for this *small* sample by visual examination.

WORKED-OUT PROBLEM 10 You want to determine the standard deviation for the restaurant meal study. For city meal costs, the standard deviation is $14.92, and you determine that the majority of meals will cost between $34.38 and $64.22 (the mean $49.30 ± $14.92). For suburban meal costs, the standard deviation is $11.38, and you determine that the majority of those meals will cost between $33.02 and $55.78 (the mean $44.40 ± $11.38).

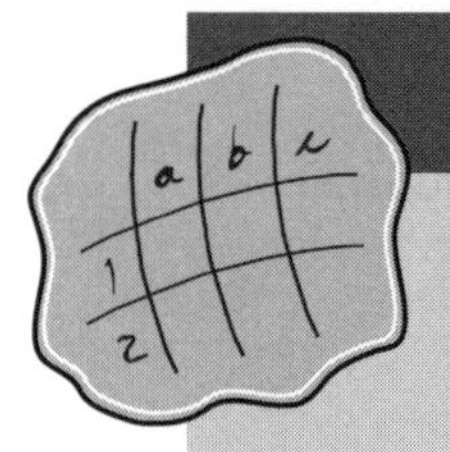

spreadsheet solution

Measures of Variation

The **Chapter 3 Descriptive** spreadsheet first mentioned on page 45 also calculates measures of variation (shown below) for the sample of get-ready times. Experiment with the spreadsheet by entering your own data in column B.

Best Practices

Enter the data for the variable being summarized in its own column.

Use the **VAR.S** and **STDEV.S** functions to calculate the variance and standard deviation for a sample. Use the **VAR.P** and **STDEV.P** functions to calculate the measures for a population.

Use the **MAX** and **MIN** functions to first identify the maximum and minimum values and then use a formula that takes the difference of those two values to calculate the range.

How-Tos

Tip FT1 in Appendix D explains how to enter the functions used in **Chapter 3 Descriptive**.

Tip ATT2 in Appendix E describes using the Analysis ToolPak as a second way to generate measures of variation.

	A	B	C	D	
1	Data		Descriptive Statistics		
13			Measures of Variation		
14			Maximum	52	=MAX(A:A)
15			Minimum	29	=MIN(A:A)
16			Range	23	=D14 - D15
17			Variance	45.82	=VAR.S(A:A)
18			Standard Deviation	6.77	=STDEV.S(A:A)

equation blackboard (optional)

interested in math?

Using symbols first introduced earlier in this chapter, you can express the sample variance and the sample standard deviation as

$$\text{Sample variance} = S^2 = \frac{\sum (X_i - \overline{X})^2}{n-1}$$

$$\text{Sample standard deviation} = S = \sqrt{\frac{\sum (X_i - \overline{X})^2}{n-1}}$$

To calculate the variance and standard deviation for population data, change the divisor from one less than the sample size (the number of data values in the sample) to the number of data values in the population, a value known as the **population size** and represented by an italicized uppercase *N*.

$$\text{Population variance} = \sigma^2 = \frac{\sum (X_i - \mu)^2}{N}$$

$$\text{Population standard deviation} = \sigma = \sqrt{\frac{\sum (X_i - \mu)^2}{N}}$$

Statisticians use the lowercase Greek letter sigma, σ, to represent the population standard deviation, replacing the uppercase italicized *S*. In statistics, symbols for population parameters are always Greek letters. (Note that the lowercase Greek letter mu, μ, which represents the *population* mean, replaces the sample mean, $\overline{X}$, in the equations for the population variance and standard deviation.)

Standard *(Z)* Score

CONCEPT The number that is the difference between a data value and the mean of the variable, divided by the standard deviation.

EXAMPLE The Z score for a particular incoming freshman's SAT score, the Z score for the get-ready time on day 4.

INTERPRETATION Z scores help you determine whether a data value is an extreme value, or *outlier*—that is, far from the mean. As a general rule, a data value's Z score that is less than –3 or greater than +3 indicates that the data value is an extreme value.

WORKED-OUT PROBLEM 11 You need to know whether any of the times from the set of ten get-ready times (see page 38) could be considered outliers. You calculate Z scores for each of those times (shown below) and compare. From these results, you learn that the greatest positive Z score was 1.83 (for the day 4 value) and the greatest negative Z score was –1.27 (for the day 8 value). Because no Z score is less than –3 or greater than +3, you conclude that none of the get-ready times can be considered extreme.

Day	Time	Time Minus the Mean	*Z* Score
1	39	–0.6	–0.09
2	29	–10.6	–1.57
3	43	3.4	0.50
4	52	12.4	1.83
5	39	–0.6	–0.09
6	44	4.4	0.65
7	40	0.4	0.06
8	31	–8.6	–1.27
9	44	4.4	0.65
10	35	–4.6	–0.68

equation blackboard (optional)

Using symbols presented earlier in this chapter, you can express the Z score as

$$Z \text{ score} = Z = \frac{X - \bar{X}}{S}$$

3.4 Shape of Distributions

Shape, a third important property of a set of numerical data, describes the pattern of the distribution of data values in a set data values. There are three possibilities: symmetric, left-skewed, or right-skewed. Shape is important as a set of data values that is too badly skewed can make certain statistical methods invalid (as explained later in this book).

Symmetrical Shape

CONCEPT A set of data values in which the mean equals the median value and each half of the curve is a mirror image of the other half of the curve.

EXAMPLE Scores on a standardized exam, actual amount of soft drink in a one-liter bottle.

Left-Skewed Shape

CONCEPT A set of data values in which the mean is less than the median value and the left tail of the distribution is longer than the right tail of the distribution. Also known as negative skew.

EXAMPLE Scores on an exam in which most students score between 70 and 100, whereas a few students score between 10 and 69.

Right-Skewed Shape

CONCEPT A set of data values in which the mean is greater than the median value and the right tail of the distribution is longer than the left tail of the distribution. Also known as positive skew.

EXAMPLE Prices of homes in a particular community, annual family income.

INTERPRETATION Right or positive skewness occurs when the set of data contains some extremely high data values (those values increase the mean). Left or negative skewness occurs when the set of data contains some extremely low values (that decrease the mean). The set of data values are symmetrical when low and high values balance each other out.

When identifying shape, you should avoid the common pitfall of thinking that the side of the histogram in which most data values cluster closely together is the direction of the skew. For example, consider the three histograms shown on the next page. Clustering in the first histogram appears toward the right of the histogram, but the pattern is properly labeled left-skewed. To see the shape more clearly, statisticians create area-under-the-curve or distribution graphs, in which a plotted, curved line represents the tops of all the bars. The equivalent distribution graphs for the three histograms are shown below the histograms. If you remember that in such graphs the longer tail points to the skewness, you will never wrongly identify the direction of the skew.

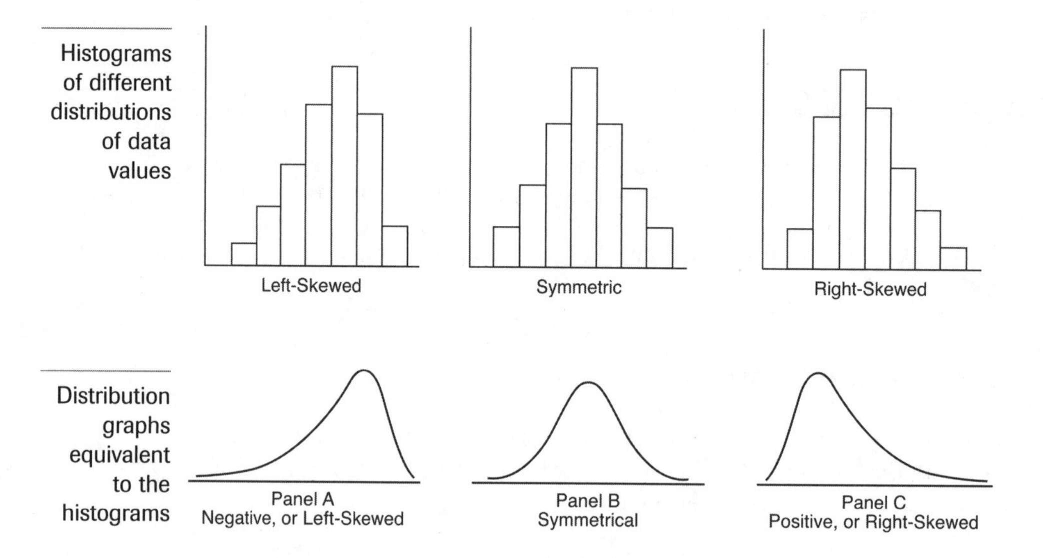

In lieu of graphing a distribution, a skewness statistic can also be calculated. For this statistic, a value of 0 means a perfectly symmetrical shape.

WORKED-OUT PROBLEM 12 You want to identify the shape of the NBA fan cost index data first presented on page 23. You examine the histogram of these data (see page 24) and determine that the distribution appears to be right-skewed because more low values than high values exist.

WORKED-OUT PROBLEM 13 The skewness for the get-ready times data is 0.086. Because this value is so close to 0, you conclude that the distribution of get-ready times around the mean is also approximately symmetric.

The Box-and-Whisker Plot

CONCEPT For a set of data values, the five numbers that correspond to the smallest value, the first quartile Q_1, the median, the third quartile Q_3, and the largest value. This plot is also known as a boxplot.

INTERPRETATION The five-number summary concisely summarizes the shape of a set of data values. This plot determines the degree of symmetry (or skewness) based on the distances that separate the five numbers. To compare these distances effectively, you can create a **box-and-whisker plot**. In this plot, the five numbers are plotted as vertical lines, interconnected so as, with some imagination, to form a "box" from which a pair of cat whiskers sprout.

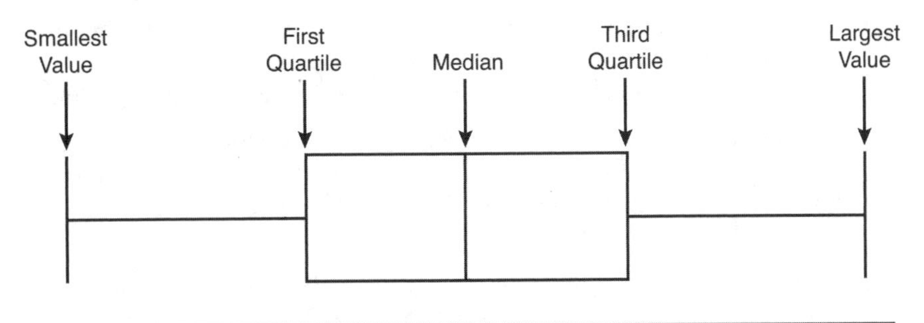

A box-and-whisker plot shows a symmetric shape for a set of data values if the following relationships are present in the plot:

- The distance from the line that represents the smallest value to the line that represents the median equals the distance from the line that represents the median to the line that represents the largest value.
- The distance from the line that represents the smallest value to the line that represents the first quartile equals the distance from the line that represents the third quartile to the line that represents the largest value.
- The distance from the line that represents the first quartile to the line that represents the median equals the distance from the line that represents the median to the line that represents the third quartile.

A box-and-whisker plot shows a right-skewed shape for a set of data values if the following relationships are present in the plot:

- The distance from the line that represents the median to the line that represents the largest value is greater than the distance from the line that represents the smallest value to the line that represents the median.
- The distance from the line that represents the third quartile to the line that represents the largest value is greater than the distance from the line that represents the smallest value to the line that represents the first quartile.

- The distance from the line that represents the first quartile to the line that represents the median is less than the distance from the line that represents the median to the line that represents the third quartile.

A box-and-whisker plot shows a left-skewed shape for a set of data values if the following relationships are present in the plot:

- The distance from the line that represents the smallest value to the line that represents the median is greater than the distance from the line that represents the median to the line that represents the largest value.
- The distance from the line that represents the smallest value to the line that represents the first quartile is greater than the distance from the line that represents the third quartile to the line that represents the largest value.
- The distance from the line that represents the first quartile to the line that represents the median is greater than the distance from the line that represents the median to the line that represents the third quartile.

WORKED-OUT PROBLEM 14 The following figure represents a box-and-whisker plot of the times to get ready in the morning:

Box-and-Whisker Plot for Get-Ready Times

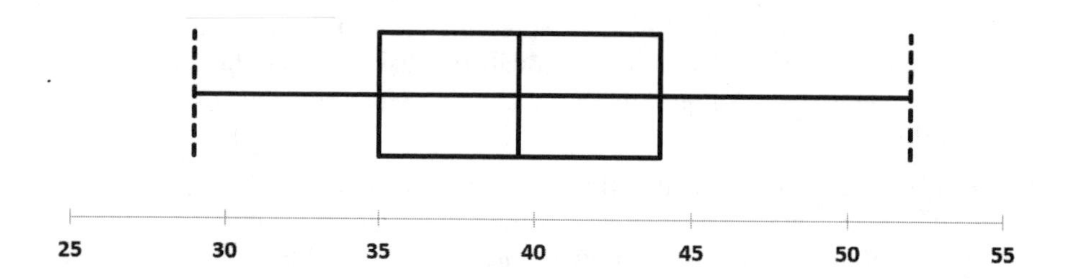

The box-and-whisker plot seems to indicate an approximately symmetric distribution of the time to get ready. The line that represents the median in the middle of the box is approximately equidistant between the ends of the box, and the length of the whiskers does not appear to be very different.

WORKED-OUT PROBLEM 15 You seek to better understand the shape of the restaurant meal cost study data used in an earlier worked-out problem. You create box-and-whisker plots for the meal cost of both the city and suburban groups.

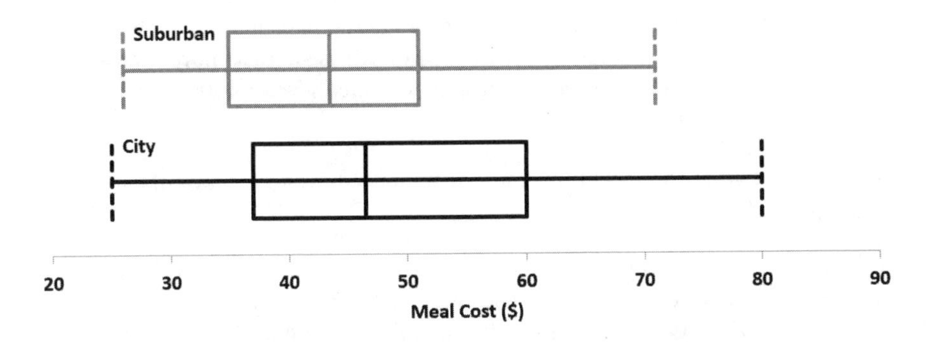

In examining the box-and-whisker plot for the city meal costs, you discover the following:

- The distance from the line that represents the smallest value ($25) to the line that represents the median ($46.50) is much less than the distance from the line that represents the median to the line that represents the highest value ($80).
- The distance from the line that represents the smallest value ($25) to the line that represents the first quartile ($36.75) is less than the distance from the line that represents the third quartile ($60.25) to the line that represents the highest value ($80).
- The distance from the line that represents the first quartile ($36.75) to the line that represents the median ($46.50) is less than the distance from the line that represents the median ($46.50) to the line that represents the third quartile ($60.25).

You conclude that the restaurant meal costs for city restaurants are right-skewed.

In examining the box-and-whisker plot for the suburban meal costs, you discover the following:

- The distance from the line that represents the smallest value ($26) to the line that represents the median ($43.50) is much less than the distance from the line that represents the median to the line that represents the highest value ($71).
- The distance from the line that represents the smallest value ($26) to the line that represents the first quartile ($34.75) is much less than the

distance from the line that represents the third quartile ($51) to the highest value ($71).

- The distance from the line that represents the first quartile ($34.75) to the line that represents the median ($43.50) is slightly more than the distance from the line that represents the median ($43.50) to the line that represents the third quartile ($51).

You conclude that the restaurant meal costs for suburban restaurants are right-skewed.

In comparing the city and suburban meal cost, you conclude that the city cost is higher than the suburban cost, because the minimum value, first quartile, median, third quartile, and maximum value are higher for the city restaurants.

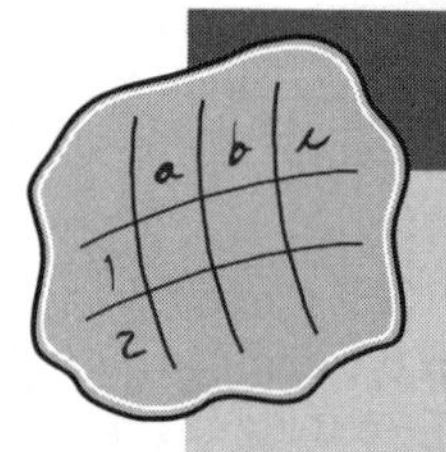

spreadsheet solution

Measures of Shape

Chapter 3 BoxWhisker Plot contains a box-and-whisker plot for the get-ready times, similar to the plot shown on page 54. Experiment with this chart by entering your own data in column A.

Best Practices

Use the **SKEW** function to calculate the skewness.

Use the **MIN**, **QUARTILE.EXC**, **MEDIAN**, and **MAX** functions to compute the five-number summary that serves as the source of the box-and-whisker plot.

How-Tos

Tip FT1 in Appendix D explains how to enter the functions used in **Chapter 3 Descriptive**.

Using Tip ATT2 in Appendix E creates results that include a measure of skewness.

Excel does not contain a box-and-whisker plot chart type. **Chapter 3 BoxWhisker Plot** uses a series of "tricked-up" line plots to create the box-and-whisker plot using the calculated five-number summary in columns C and D and the formulas and data hidden behind the plot in columns F and G.

Important Equations

Mean: (3.1) $\overline{X} = \frac{\sum X_i}{n}$

Median: (3.2) $\text{Median} = \frac{n+1}{2} th \text{ ranked value}$

First Quartile Q_1: (3.3) $Q_1 = \frac{n+1}{4} th \text{ ranked value}$

Third Quartile Q_3: (3.4) $Q_3 = \frac{3(n+1)}{4} th \text{ ranked value}$

Range: (3.5) Range = *largest value – smallest value*

Sample Variance: (3.6) $S^2 = \frac{\sum (X_i - \overline{X})^2}{n-1}$

Sample Standard Deviation: (3.7) $S = \sqrt{\frac{\sum (X_i - \overline{X})^2}{n-1}}$

Population Variance: (3.8) $\sigma^2 = \frac{\sum (X_i - \mu)^2}{N}$

Population Standard Deviation: (3.9) $\sigma = \sqrt{\frac{\sum (X_i - \mu)^2}{N}}$

Z Scores: (3.10) $Z = \frac{X - \overline{X}}{S}$

One-Minute Summary

The properties of central tendency, variation, and shape enable you to describe a set of data values for a numerical variable.

Numerical Descriptive Measures

- Central tendency
  - Mean
  - Median
  - Mode
- Variation
  - Range
  - Variance
  - Standard deviation
  - Z scores
- Shape
  - Skew statistic
  - Five-number summary
  - Box-and-whisker plot

Test Yourself

Short Answers

1. Which of the following statistics are measures of central tendency?
   (a) median
   (b) range
   (c) standard deviation
   (d) all of these
   (e) none of these
2. Which of the following statistics is not a measure of central tendency?
   (a) mean
   (b) median
   (c) mode
   (d) range
3. Which of the following statements about the median is not true?
   (a) It is less affected by extreme values than the mean.
   (b) It is a measure of central tendency.
   (c) It is equal to the range.
   (d) It is equal to the mode in bell-shaped "normal" distributions.

4. Which of the following statements about the mean is not true?
   (a) It is more affected by extreme values than the median.
   (b) It is a measure of central tendency.
   (c) It is equal to the median in skewed distributions.
   (d) It is equal to the median in symmetric distributions.

5. Which of the following measures of variability is dependent on every value in a set of data?
   (a) range
   (b) standard deviation
   (c) each of these
   (d) neither of these

6. Which of the following statistics cannot be determined from a box-and-whisker plot?
   (a) standard deviation
   (b) median
   (c) range
   (d) the first quartile

7. In a symmetric distribution:
   (a) the median equals the mean
   (b) the mean is less than the median
   (c) the mean is greater than the median
   (d) the median is less than the mode

8. The shape of a distribution is given by the:
   (a) mean
   (b) first quartile
   (c) skewness
   (d) variance

9. In a five-number summary, the following is not included:
   (a) median
   (b) third quartile
   (c) mean
   (d) minimum (smallest) value

10. In a right-skewed distribution:
   (a) the median equals the mean
   (b) the mean is less than the median
   (c) the mean is greater than the median
   (d) the median equals the mode

Answer True or False:

11. In a box-and-whisker plot, the box portion represents the data between the first and third quartile values.
12. The line drawn within the box of the box-and-whisker plot represents the mean.

Fill in the blanks:

13. The _______ is found as the middle value in a set of values placed in order from lowest to highest for an odd-sized sample of numerical data.
14. The standard deviation is a measure of _______.
15. If all the values in a data set are the same, the standard deviation will be _______.
16. A distribution that is negative-skewed is also called _______-skewed.
17. If each half of a distribution is a mirror image of the other half of the distribution, the distribution is called _______.
18. The median is a measure of _______.

19, 20, 21. The three characteristics that describe a set of numerical data are _______, _______, and _______.

For Questions 22 through 30, the number of days absent by a sample of nine students during a semester was as follows:

9 1 1 10 7 11 5 8 2

22. The mean is equal to _______.
23. The median is equal to _______.
24. The mode is equal to _______.
25. The first quartile is equal to _______.
26. The third quartile is equal to _______.
27. The range is equal to _______.
28. The variance is approximately equal to _______.
29. The standard deviation is approximately equal to _______.
30. The data are:
    (a) right-skewed
    (b) left-skewed
    (c) symmetrical

31. In a left-skewed distribution:
    (a) the median equals the mean
    (b) the mean is less than the median
    (c) the mean is greater than the median
    (d) the median equals the mode

32. Which of the statements about the standard deviation is true?
    (a) It is a measure of variation around the mean.
    (b) It is the square of the variance.
    (c) It is a measure of variation around the median.
    (d) It is a measure of central tendency.

33. The smallest possible value of the standard deviation is _______.

Answers to Test Yourself Short Answers

1. a
2. d
3. c
4. c
5. b
6. a
7. a
8. c
9. c
10. c
11. True
12. False
13. median
14. variation
15. 0
16. left
17. symmetric
18. central tendency
19. central tendency
20. variation
21. shape
22. 6
23. 7
24. 1
25. 1.5
26. 9.5
27. 10
28. 15.25
29. 3.91
30. b
31. b
32. a
33. 0

FoodPrices

Problems

1. How do prices at online grocers compare to prices at supermarkets? A market basket of 14 items was purchased with the following results:

Company	Total Cost	Type
Morton Williams Store	72.77	Supermarket
Peapod (New York)	72.95	Online
FreshDirect	75.13	Online
Safeway Store	58.16	Supermarket
Safeway.com	75.85	Online
AmazonFresh	62.13	Online
Walmart.com	52.70	Online
Instacart	72.19	Online
Kroger Store	57.00	Supermarket
Peapod (Indianapolis)	70.57	Online

Source: Data extracted from G. A. Fowler, "Price Check: Do Online Grocers Beat Supermarkets?" *The Wall Street Journal*, 8 January 2014, p. D1–D2.

For supermarkets and online grocers separately,

(a) Compute the mean and median.

(b) Compute the variance, standard deviation, and range.

(c) Based on the results of (a) and (b), what conclusions can you reach concerning the cost of a basket of products between online grocers and supermarkets?

2. The following data represents the overall miles per gallon (MPG) of 2014 midsized sedans:

38 26 30 26 25 27 24 22 27 32 39
26 24 24 23 24 25 31 26 37 22 33

Sedans

Source: Data extracted from "Which Car Is Right For You," *Consumer Reports*, April 2014, p. 40–41.

(a) Compute the mean and median.

(b) Compute the first quartile and the third quartile.

(c) Compute the variance, standard deviation, and range.

(d) Construct a box-and-whisker plot.

(e) Are the data skewed? If so, how?

(f) Based on the results of (a) through (d), what conclusions can you reach concerning the miles per gallon of midsized sedans?

NBACost

3. As player salaries have increased, the cost of attending NBA professional basketball games has increased dramatically. The following data represents the Fan Cost Index, which is the cost of four tickets, two beers, four soft drinks, four hot dogs, two game programs, two caps, and the parking fee for one car at each of the 30 NBA arenas:

240.04 434.96 382.00 203.06 456.60 271.74 321.18 319.10 262.40 324.08
336.05 227.36 395.20 542.00 212.16 472.20 309.30 273.98 208.48 659.92
295.40 263.10 266.40 344.92 308.18 268.28 338.00 321.63 280.98 249.22

Source: Data extracted from "NBA Fan Cost Experience," **bit.ly/1nnu9rf**.

   (a) Compute the mean and median.
   (b) Compute the first quartile and the third quartile.
   (c) Compute the variance, standard deviation, and range.
   (d) Construct a box-and-whisker plot.
   (e) Are the data skewed? If so, how?
   (f) Based on the results of (a) through (d), what conclusions can you reach concerning the Fan Cost Index of NBA games?

DomesticBeer

4. The file **DomesticBeer** contains the percent alcohol, number of calories per 12 ounces, and number of carbohydrates (in grams) per 12 ounces for 152 of the best-selling domestic beers in the United States. (Data extracted from **www.beer100.com/beercalories.htm**, 20 March 2013.)

   For each variable:
   (a) Compute the mean and median.
   (b) Compute the first quartile and the third quartile.
   (c) Compute the variance, standard deviation, and range.
   (d) Construct a box-and-whisker plot.
   (e) Are the data skewed? If so, how?
   (f) Based on the results of (a) through (d), what conclusions can you reach concerning the percentage alcohol, number of calories per 12 ounces, and number of carbohydrates (in grams) per 12 ounces?

Answers to Problems

1. (a) Mean = \$62.64 for supermarkets and \$68.79 for online grocers, median = \$58.16 for supermarkets and \$72.19 for online grocers, (b) Variance = \$77.2484 for supermarkets and \$70.8861 for online grocers, standard deviation = \$8.79 for supermarkets and \$8.42 for online grocers, range = \$15.77 for supermarkets and \$23.15 for online grocers (c) The mean for supermarkets is slightly more than the median, so the data are slightly right-skewed. The mean for online grocers is slightly less than the median, so the data are slightly left-skewed.

(d) The mean cost for supermarkets is slightly less than the mean cost for online grocers. The median cost for supermarkets is substantially less than the median cost for online grocers. The standard deviation is slightly lower and the range is somewhat lower for supermarkets than for online grocers. Note, however, that the sample sizes used here are extremely small especially for the supermarkets.

2. (a) Mean = 27.7727, median = 26
(b) Q_1 = 24 Q_3 = 31.25
(c) Variance = 26.2792, standard deviation = 5.1263, range = 17
(e) The mean is higher than the median. The difference between the largest value and Q_3 is 7.75, while the difference between Q_1 and the smallest value is 2, so the data are right-skewed.
(f) The mean miles per gallon of midsized sedans is 27.7727. Miles per gallon are greater than or equal to 26 for half the midsized sedans and less than or equal to 26 for half the midsized sedans. The average scatter around the mean miles per gallon is 5.1263. The difference between the highest and lowest miles per gallon is 17. The miles per gallon are right-skewed because of the several extremely high values for the hybrid vehicles.

3. (a) Mean = $326.26., median = $308.74
(b) Q_1 = $262.93 Q_3 = $354.19
(c) Variance = $10,554.2976, standard deviation = $102.7341, range = $456.86
(e) The mean is larger than the median. The difference between the largest value and Q_3 is $305.73, while the difference between Q_1 and the smallest value is $59.87, so the data are right-skewed.
(f) The mean fan cost index is $326.26 whereas the fan cost index is below $308.74 for half the teams and above $308.74 for half the teams. The average scatter of the fan cost index around the mean is $102.7341. The difference between the highest and the lowest fan cost index is $456.86. The fan cost index is right skewed as the fan cost index is extremely high for several teams especially the Los Angeles Lakers and New York Knicks.

4. (a) The means are percent alcohol, 5.236; calories, 154.309; and carbohydrates, 11.964.
The medians are percent alcohol, 4.9; calories, 150; and carbohydrates, 12.055.
(b) The first quartiles are percent alcohol, 4.4; calories, 129.25; and carbohydrates, 8.375.
The third quartiles are percent alcohol, 5.6; calories, 166; and carbohydrates, 14.45.
(c) The variances are percent alcohol, 2.045; calories, 1,987.473; and carbohydrates, 24.227.
The standard deviations are percent alcohol, 1.430; calories, 44.581; and carbohydrates, 4.922.

The ranges are percent alcohol, 11.1; calories, 275; and carbohydrates, 30.2.
(d) Box-and-whisker plot not shown.
(e) The percent alcohol is right-skewed from the box-and-whisker plot with a mean of 5.24%. Half of the beers have less than 4.9% alcohol. The middle 50% of the beers have alcohol content spread over a range of 1.2%. The highest alcohol content is 11.5%, and the lowest is 0.4%. The average scatter of percent alcohol around the mean is 1.4301%. The number of calories is right-skewed from the box-and-whisker plot with a mean of 154.309. Half of the beers have calories below 150. The middle 50% of the beers have calories spread over a range of 36.75. The highest number of calories is 330, and the lowest is 55. The average scatter of calories around the mean is 44.581. The number of carbohydrates is right-skewed from the box-and-whisker plot with a mean of 11.964, which is slightly lower than the median, at 12.055. Half of the beers have carbohydrates below 12.055. The middle 50% of the beers have carbohydrates spread over a range of 6.075. The highest number of carbohydrates is 32.1, and the lowest is 1.9. The average scatter of carbohydrates around the mean is 4.9221.

References

1. Berenson, M. L., D. M. Levine, and K. A. Szabat. *Basic Business Statistics: Concepts and Applications, Thirteenth Edition*. Upper Saddle River, NJ: Pearson Education, 2015.
2. Levine, D. M., D. Stephan, and K. A. Szabat, *Statistics for Managers Using Microsoft Excel, Seventh Edition*. Upper Saddle River, NJ: Pearson Education, 2014.
3. Microsoft Excel 2013. Redmond, WA: Microsoft Corporation, 2012.

Probability

Probability theory serves as one of the building blocks for inferential statistics. This chapter reviews the probability concepts necessary to comprehend the statistical methods discussed in later chapters. If you are already familiar with probability, you should skim this chapter to learn the vocabulary that this book uses to discuss probability concepts.

4.1 Events

Events underlie all discussions about probability. Before you can define probability in statistical terms, you need to understand the meaning of an event.

Event

CONCEPT An outcome of an experiment or survey.

EXAMPLES Rolling a die and turning up six dots, an individual who votes for the incumbent candidate in an election, someone using a social media website.

INTERPRETATION Recall from Chapter 1 (see page 5) that performing experiments or conducting surveys are two important types of data sources. When discussing probability, many statisticians use the word *experiment* broadly to include surveys, so you can use the shorter definition "an outcome of an experiment" if you understand this broader usage of *experiment*. Likewise, as you read this chapter and encounter the word *experiment*, you should use the broader meaning.

Elementary Event

CONCEPT An outcome that satisfies only one criterion.

EXAMPLES A red card from a standard deck of cards, a voter who selected the Democratic candidate, a person who owns a 4G LTE smartphone.

INTERPRETATION Elementary events are distinguished from joint events, which meet two or more criteria.

Joint Event

CONCEPT An outcome that satisfies two or more criteria.

EXAMPLES A red ace from a standard deck of cards, a voter who voted for the Democratic candidate for president and the Democrat candidate for U.S. senator, a female who owns a 4G LTE smartphone.

INTERPRETATION Joint events are distinguished from elementary events which only meet one criterion.

4.2 More Definitions

Using the concept of **event**, you can define three more basic terms of probability.

Random Variable

CONCEPT A variable whose numerical values represent the events of an experiment.

EXAMPLES The number of text messages sent by a certain person in a 24-hour period, the scores of students on a standardized exam, the preferences of consumers for different brands of automobiles.

INTERPRETATION You use the phrase **random variable** to refer to a variable that has no data values until an experimental trial is performed or a survey question is asked and answered. Random variables are either discrete, in which the possible numerical values are a set of integers (or coded values in the case of categorical data); or continuous, in which any value is possible within a specific range.

Probability

CONCEPT A number that represents the chance that a particular event will occur for a random variable.

EXAMPLES Odds of winning a lottery, chance of rolling a seven when rolling two dice, likelihood of an incumbent winning reelection, chance that an individual will own a 4G LTE smartphone.

INTERPRETATION Probability determines the likelihood that a random variable will be assigned a specific value. Probability considers things that might occur in the future, and its forward-looking nature provides a bridge to inferential statistics.

Probabilities can be developed for an elementary event of a random variable or any group of joint events. For example, when rolling a standard six-sided die (see illustration), six possible elementary events correspond to the six faces of the die that contain either one, two, three, four, five, or six dots. "Rolling a die and turning up an even number of dots" is an example of an event formed from three elementary events (rolling a two, four, or six).

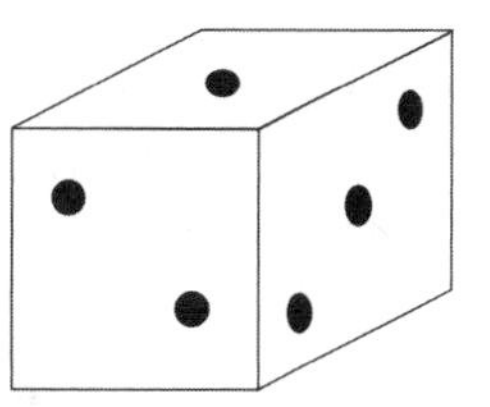

Probabilities are formally stated as decimal numbers in the range of 0 to 1. A probability of 0 indicates an event that never occurs (such an event is known as a **null event**). A probability of 1 indicates a **certain event**, an event that must occur. For example, when you roll a die, getting seven dots is a null event, because it can never happen, and getting six or fewer dots is a certain event, because you will always end up with a face that has six or fewer dots. Probabilities can also be stated informally as the "percentage chance of (something)" or as quoted odds, such as a "50-50 chance."

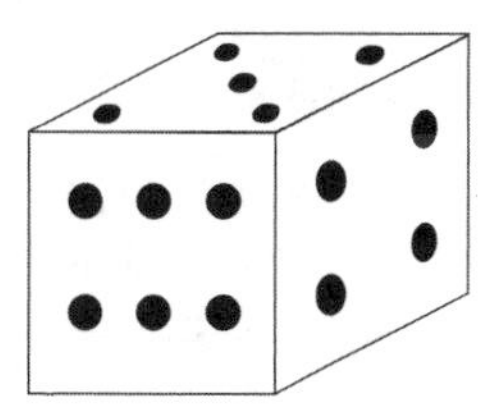

Collectively Exhaustive Events

CONCEPT A set of events that includes all the possible events.

EXAMPLES Heads and tails in the toss of a coin, male and female, all six faces of a die.

INTERPRETATION When you have a set of collectively exhaustive events, one of them must occur. The coin must land on either heads or tails; the person must be male or female; the die must end with a face that has six or fewer dots. The sum of the individual probabilities associated with a set of collectively exhaustive events is always 1.

4.3 Some Rules of Probability

A set of rules govern the calculation of the probabilities of elementary and joint events.

RULE 1 The probability of an event must be between 0 and 1. The smallest possible probability value is 0. You cannot have a negative probability. The largest possible probability value is 1.0. You cannot have a probability greater than 1.0.

EXAMPLE In the case of a die, the event of getting a face of seven has a probability of 0, because this event cannot occur. The event of getting a face with fewer than seven dots has a probability of 1.0, because it is certain that one of the elementary events of one, two, three, four, five, or six dots must occur.

RULE 2 The event that *A* does not occur is called ***A* complement** or simply **not *A***, and is given the symbol *A'*. If *P(A)* represents the probability of event *A* occurring, then 1 – *P(A)* represents the probability of event A not occurring.

EXAMPLE In the case of a die, the complement of getting the face that contains three dots is not getting the face that contains three dots. Because the probability of getting the face containing three dots is 1/6, the probability of not getting the face that contains three dots is (1 –1/6) = 5/6 or 0.833.

RULE 3 If two events *A* and *B* are **mutually exclusive**, the probability of both events *A* and *B* occurring is 0. This means that the two events cannot occur at the same time.

EXAMPLE On a single roll of a die, you cannot get a die that has a face with three dots and also have four dots because such elementary events are mutually exclusive. Either three dots can occur or four dots can occur, but not both.

RULE 4 If two events *A* and *B* are mutually exclusive, the probability of either event *A* or event *B* occurring is the sum of their separate probabilities.

EXAMPLE The probability of rolling a die and getting either a face with three dots or a face with four dots is 1/3 or 0.333, because these two events are mutually exclusive. Therefore, the probability of 1/3 is the sum of the probability of rolling a three (1/6) and the probability of rolling a four (1/6), which is 1/3.

INTERPRETATION You can extend this addition rule for mutually exclusive events to situations in which more than two events exist. In the case of rolling a die, the probability of turning up an even face (two, four, or six dots) is 0.50, the sum of 1/6 and 1/6 and 1/6 (3/6, or 0.50).

RULE 5 If events in a set are mutually exclusive and collectively exhaustive, the sum of their probabilities must add up to 1.0.

EXAMPLE The events of a turning up a face with an even number of dots and turning up a face with an odd number of dots are mutually exclusive and collectively exhaustive. They are mutually exclusive, because even and odd cannot occur simultaneously on a single roll of a die. They are also collectively exhaustive, because either even or odd must occur on a particular roll. Therefore, for a single die, the probability of turning up a face with an even or odd face is the sum of the probability of turning up an even face plus the probability of turning up an odd face, or 1.0, as follows:

$$\begin{aligned} P\text{ (even or odd face)} &= P\text{ (even face)} + P\text{ (odd face)} \\ &= \frac{3}{6} + \frac{3}{6} \\ &= \frac{6}{6} = 1 \end{aligned}$$

RULE 6 If two events *A* and *B* are not mutually exclusive, the probability of either event *A* or event *B* occurring is the sum of their separate probabilities minus the probability of their simultaneous occurrence (the joint probability).

EXAMPLE For rolling a single die, turning up a face with an even number of dots is not mutually exclusive with turning up a face with fewer than five dots, because both events include these (two) elementary events: turning up the face with two dots and turning up the face with four dots. To determine the probability of these two events, you add the probability of having a face with an even number of dots (3/6) to the probability of having a face with fewer than five dots (4/6) and then subtract the joint probability of simultaneously having a face with an even number of dots and having a face with fewer than five dots (2/6). You can express this as follows:

P (even face *or* face with fewer than five dots) =

P (even face) + *P* (face with fewer than five dots) –
P (even face *and* face with fewer than five dots)

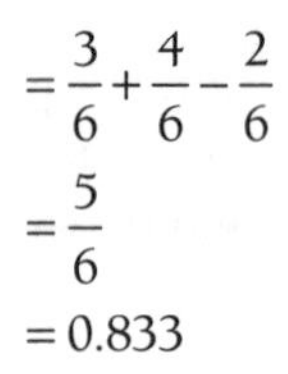

$$= \frac{3}{6} + \frac{4}{6} - \frac{2}{6}$$
$$= \frac{5}{6}$$
$$= 0.833$$

INTERPRETATION This rule requires that you subtract the joint probability, because that probability has already been included twice (in the first event and in the second event). Because the joint probability has been "double counted," you must subtract it to compute the correct result.

RULE 7 If two events *A* and *B* are **independent**, the probability of both events *A* and *B* occurring is equal to the product of their individual probabilities. Two events are independent if the occurrence of one event in no way affects the probability of the second event.

EXAMPLE When rolling a die, each roll of the die is an independent event, because no roll can affect another (although gamblers who play dice games sometimes would like to think otherwise). Therefore, to determine the probability that two consecutive rolls both turn up the face with five dots, you multiply the probability of turning up that face on roll one (1/6) by the probably of turning up that face on roll two (also 1/6). You can express this as follows:

P (face with five dots on roll one and face with five dots on roll two) =

P (face with five dots on roll one) × *P* (face with five dots on roll two)

$$= \frac{1}{6} \times \frac{1}{6}$$
$$= \frac{1}{36} = 0.028$$

RULE 8 If two events *A* and *B* are not independent, the probability of both events *A* and *B* occurring is the product of the probability of event *A* multiplied by the probability of event *B* occurring, given that event A has occurred.

EXAMPLE During the taping of a television game show, contestants are randomly selected from the audience watching the show. After a particular person has been chosen, he or she does not return to the audience and cannot be chosen again, therefore making this a case in which the two events are not independent.

If the audience consists of 30 women and 20 men (50 people), what is the probability that the first two contestants chosen are male? The probability that the first contestant is male is simply 20/50 or 0.40. However, the probability that the second contestant is male is not 20/50, because when the second selection is made, the eligible audience now has only 19 males and 49 people because the first male selected cannot be selected again. Therefore, the probability that the second selection is male is 19/49 or 0.388, rounded. This means that the probability that the first two contestants are male is 0.155, as follows:

$$P\text{ (male selection first and male selection second)} =$$
$$P\text{ (male selection first)} \times P\text{ (male selection second)}$$

$$= \frac{20}{50} \times \frac{19}{49}$$
$$= \frac{380}{2,450} = 0.155$$

4.4 Assigning Probabilities

Three different approaches exist for assigning probabilities to the events of a random variable: the classical approach, the empirical approach, and the subjective approach.

Classical Approach

CONCEPT Assigning probabilities based on prior knowledge of the process involved.

EXAMPLE Rolling a die and assigning the probability of turning up the face with three dots.

INTERPRETATION Classical probability often assumes that all elementary events are equally likely to occur. When this is true, the probability that a particular event will occur is defined by the number of ways the event can occur divided by the total number of elementary events. For example, when you roll a die, the probability of getting the face with three dots is 1/6 because six elementary events are associated with rolling a die. Thus, you can expect that 1,000 out of 6,000 rolls of a die would turn up the face with three dots.

Empirical Approach

CONCEPT Assigning probabilities based on frequencies obtained from empirically observed data.

EXAMPLE Probabilities determined by polling or marketing surveys.

INTERPRETATION The empirical approach does not use theoretical reasoning or assumed knowledge of a process to assign probabilities. Similar to the classical approach when all elementary events are equally likely, the empirical probability can be calculated by dividing the number of ways *A* can occur by the total number of elementary events. For example, if a poll of 500 registered voters reveals that 275 are likely to vote in the next election, you can assign the empirical probability of 0.55 (275 divided by 500).

Subjective Approach

CONCEPT Assign probabilities based on expert opinions or other subjective methods such as "gut" feelings or hunches.

EXAMPLE Commentators stating the odds that a political candidate will win an election or that a sports team will win a championship, a financial analyst stating the chance that a stock will increase in value by a certain amount in the next year.

INTERPRETATION In this approach, you use your own intuition and knowledge or experience to judge the likeliest outcomes. You use the subjective approach when either the number of elementary events or actual data are not available for the calculation of relative frequencies. Because of the subjectivity, different individuals might assign different probabilities to the same event.

One-Minute Summary

Foundation Concepts

- Rules of probability
- Assigning probabilities

Test Yourself

Short Answers

1. If two events are collectively exhaustive, what is the probability that one or the other occurs?
   (a) 0
   (b) 0.50
   (c) 1.00
   (d) Cannot be determined from the information given

2. If two events are collectively exhaustive, what is the probability that both occur at the same time?
   (a) 0
   (b) 0.50
   (c) 1.00
   (d) Cannot be determined from the information given
3. If two events are mutually exclusive, what is the probability that both occur at the same time?
   (a) 0
   (b) 0.50
   (c) 1.00
   (d) Cannot be determined from the information given
4. If the outcome of event *A* is not affected by event *B*, then events *A* and *B* are said to be:
   (a) mutually exclusive
   (b) independent
   (c) collectively exhaustive
   (d) dependent

Use the following problem description when answering Questions 5 through 9:

A survey is taken among customers of a coffee shop to determine preference for regular or decaffeinated coffee. Of 200 respondents selected, 125 were male and 75 were female. 120 preferred regular coffee and 80 preferred decaffeinated coffee. Of the males, 85 preferred regular coffee.

5. The probability that a randomly selected individual is a male is:
   (a) 125/200
   (b) 75/200
   (c) 120/200
   (d) 200/200
6. The probability that a randomly selected individual prefers regular or decaffeinated coffee is:
   (a) 0/200
   (b) 125/200
   (c) 75/200
   (d) 200/200
7. Suppose that two individuals are randomly selected. The probability that both prefer regular coffee is:
   (a) (120/200)(120/200)
   (b) (120/200)
   (c) (120/200)(119/199)
   (d) (85/200)

8. The probability that a randomly selected individual prefers regular coffee is:
   (a) 0/200
   (b) 120/200
   (c) 75/200
   (d) 200/200
9. The probability that a randomly selected individual prefers regular coffee *or* is a male is:
   (a) 0/200
   (b) 125/200
   (c) 160/200
   (d) 200/200
10. The smallest possible value for a probability is __________.
11. The largest possible value for a probability is __________.
12. If two events are __________, they cannot occur at the same time.
13. If two events are __________, the probability that both events occur is the product of their individual probabilities.
14. In the __________ probability approach, probabilities are based on frequencies obtained from surveys.
15. In the __________ probability approach, probabilities can vary depending on the individual assigning them.

Answers to Test Yourself Short Answers

1. c
2. d
3. a
4. b
5. a
6. d
7. c
8. b
9. c
10. 0
11. 1
12. mutually exclusive
13. independent
14. empirical
15. subjective

Problems

1. Businesses use a method called *A/B testing* to test different web page designs to see if one design is more effective than another. For one company, designers were interested in the effect of modifying the call-to-action button on the home page. Every visitor to the company's home page was randomly shown either the original call-to-action

button (the control) or the new variation. Designers measured success by the download rate: the number of people who downloaded the file divided by the number of people who saw that particular call-to-action button. Results of the experiment yielded the following:

		Call-to-Action Button		
		Original	New	Total
Download	Yes	351	451	802
	No	3,291	3,105	6,396
	Total	3,642	3,556	7,198

If a respondent is selected at random, what is the probability that he or she:

(a) Downloaded the file?

(b) Downloaded the file *and* used the new call-to-action button?

(c) Downloaded the file *or* used the new call-to-action button?

(d) Suppose two respondents who used the original call-to-action button were selected. What is the probability that both downloaded the file?

2. A survey of 1,085 adults asked, "Do you enjoy shopping for clothing for yourself?" The results (data extracted from "Split decision on clothes shopping," *USA Today*, 28 January 2011, p. 1B) indicated that 51% of the females enjoyed shopping for clothing for themselves, compared to 44% of the males. The sample sizes of males and females was not provided. Suppose that the results were as shown in the following table:

		Gender		
		Male	Female	Total
Enjoyed Shopping	Yes	238	276	514
	No	304	267	571
	Total	542	543	1,085

What is the probability that a respondent chosen at random:

(a) Enjoys shopping for clothing for himself or herself?

(b) Is a female *and* enjoys shopping for clothing for herself?

(c) Is a female or a person who enjoys shopping for clothing?

(d) Is a male or a female?

3. In 41 of the 63 years from 1950 through 2013, the S&P 500 finished higher after the first five days of trading. In 36 of those 41 years, the S&P 500 finished higher for the year (in 2011, there was virtually no change). Is a good first week a good omen for the upcoming year? The

following table gives the first-week and annual performance over this 63-year period:

S&P 500's Annual Performance

		Year	
		Higher	**Lower**
First Five Days	**Higher**	36	5
	Lower	11	11

If a year is selected at random, what is the probability that
(a) the S&P 500 finished higher for the year?
(b) the S&P 500 finished higher after the first five days of trading?
(c) the S&P 500 finished higher after the first five days of trading *and* the S&P 500 finished higher for the year?
(d) the S&P 500 finished higher after the first five days of trading *or* the S&P 500 finished higher for the year?
(e) Given that the S&P 500 finished higher after the first five days of trading, what is the probability that it finished higher for the year?

Answers to Test Yourself Problems

1. (a) 802/7,198 = 0.1114
(b) 451/7,198 = 0.0627
(c) 3,556/7,198 + 802/7,198 – 451/7,198 = 3,907/7,198 = 0.5428
(d) (351/3,642)(350/3,641) = 0.0093
2. (a) 514/1,085 = 0.4737
(b) 276/1,085 = 0.2544
(c) 781/1,085 = 0.7198
(d) 542/1,085 + 543/1,085 = 1,085/1,085 = 1.0
3. (a) 47/63 = 0.7460
(b) 41/63 = 0.6508
(c) 36/63 = 0.5714
(d) 52/63 = 0.8254
(e) 36/41 = 0.8780

References

1. Berenson, M. L., D. M. Levine, and K. A. Szabat. *Basic Business Statistics: Concepts and Applications, Thirteenth Edition*. Upper Saddle River, NJ: Pearson Education, 2015.
2. Levine, D. M., D. Stephan, and K. A. Szabat. *Statistics for Managers Using Microsoft Excel, Seventh Edition*. Upper Saddle River, NJ: Pearson Education, 2014.

CHAPTER

5

Probability Distributions

In Chapter 4, you learned to use the rules of probability to calculate the chance that a particular event would occur. In many situations, you can use specific probability models to estimate the probability that particular events would occur.

5.1 Probability Distributions for Discrete Variables

A probability distribution for a variable summarizes or models the probabilities associated with the events for that variable. The form the distribution takes depends on whether the variable is discrete or continuous.

This section reviews probability distributions for discrete variables and statistics related to these distributions.

Discrete Probability Distribution

CONCEPT A listing of all possible distinct (elementary) events for a variable and their probabilities of occurrence.

EXAMPLE See WORKED-OUT PROBLEM 1.

INTERPRETATION In a probability distribution, the sum of the probabilities of all the events always equals 1. This is a way of saying that the (elementary) events listed are always collectively exhaustive; that is, that one of them must occur. Although you can use a table of outcomes to develop a probability distribution (see WORKED-OUT PROBLEM 2 on page 81), you can also calculate probabilities for certain types of variables by using a formula that mathematically models the distribution.

WORKED-OUT PROBLEM 1 You want to determine the probability of getting 0, 1, 2, or 3 heads when you toss a fair coin (one with an equal probability of a head or a tail) three times in a row. Because getting 0, 1, 2, or 3 heads represent all possible distinct outcomes, you form a table of all possible outcomes (eight) of tossing a fair coin three times as follows.

Outcome	First Toss	Second Toss	Third Toss
1	Head	Head	Head
2	Head	Head	Tail
3	Head	Tail	Head
4	Head	Tail	Tail
5	Tail	Head	Head
6	Tail	Head	Tail
7	Tail	Tail	Head
8	Tail	Tail	Tail

From this table of all eight possible outcomes, you can form the summary table shown in Table 5.1.

From this probability distribution, you can determine that the probability of tossing three heads in a row is 0.125 and that the sum of the probabilities is 1.0, as it should be for a distribution of a discrete variable.

TABLE 5.1

Probability Distribution for Tossing a Fair Coin Three Times

Number of Heads	Number of Outcomes with That Number of Heads	Probability
0	1	1/8 = 0.125
1	3	3/8 = 0.375
2	3	3/8 = 0.375
3	1	1/8 = 0.125

Another way to compute these probabilities is to extend Rule 7 on page 72, the multiplication rule, to three events (or tosses). To get the probability of three heads, which is equal to 1/8 or 0.125 using Rule 7, you have:

$$P(H_1 \text{ and } H_2 \text{ and } H_3) = P(H_1) \times P(H_2) \times P(H_3)$$

Because the probability of heads in each toss is 0.5:

$$P(H_1 \text{ and } H_2 \text{ and } H_3) = (0.5)\ (0.5)\ (0.5)$$
$$P(H_1 \text{ and } H_2 \text{ and } H_3) = 0.125$$

The Expected Value of a Variable

CONCEPT The sum of the products formed by multiplying each possible event in a discrete probability distribution by its corresponding probability.

INTERPRETATION The expected value tells you the value of the variable that you could expect in the "long run"; that is, after many experimental trials. The expected value of a variable is also the mean (μ) of a variable.

WORKED-OUT PROBLEM 2 If there are three tosses of a coin (refer to Table 5.1), you can calculate the expected value of the number of heads as shown in Table 5.2.

Expected or Mean Value = Sum of [each value × the probability of each value]

$$\text{Expected or Mean Value} = \mu = (0)(0.125) + (1)(0.375) + (2)(0.375) + (3)(0.125)$$
$$= 0 + 0.375 + 0.750 + 0.375 = 1.50$$

TABLE 5.2

Computing the Expected Value or Mean of a Probability Distribution

Number of Heads	Probability	(Number of Heads) × (Probability)
0	0.125	(0) × (0.125) = 0
1	0.375	(1) × (0.375) = 0.375
2	0.375	(2) × (0.375) = 0.75
3	0.125	(3) × (0.125) = 0.125
		Expected or Mean Value = 1.50

Notice that in this example, the mean or expected value of the number of heads is 1.5, a value for the number of heads that is impossible. The mean of 1.5 heads tells you that, in the long run, if you toss three fair coins many times, the mean number of heads you can expect is 1.5.

Standard Deviation of a Variable (σ)

CONCEPT The measure of variation around the expected value of a variable. You calculate this by first multiplying the squared difference between each value and the expected value by its corresponding probability. You then sum these products and then take the square root of that sum.

WORKED-OUT PROBLEM 3 If there are three tosses of a coin (refer to Table 5.1), you can calculate the variance and standard deviation of the number of heads, as shown in Table 5.3.

TABLE 5.3

Computing the Variance and Standard Deviation of a Probability Distribution

Number of Heads	Probability	(Number of Heads – Mean Number of Heads)^2 × (Probability)
0	0.125	$(0 - 1.5)^2 \times (0.125) = 2.25 \times (0.125) = 0.28125$
1	0.375	$(1 - 1.5)^2 \times (0.375) = 0.25 \times (0.375) = 0.09375$
2	0.375	$(2 - 1.5)^2 \times (0.375) = 0.25 \times (0.375) = 0.09375$
3	0.125	$(3 - 1.5)^2 \times (0.125) = 2.25 \times (0.125) = 0.28125$
		Total (Variance) = 0.75

σ = Square root of [Sum of (Squared differences between each value and the mean) × (Probability of the value)]

$$\sigma = \sqrt{(0-1.5)^2(0.125)+(1-1.5)^2(0.375)+(2-1.5)^2(0.375)+(3-1.5)^2(0.125)}$$
$$= \sqrt{2.25(0.125)+0.25(0.375)+0.25(0.375)+2.25(0.125)}$$
$$= \sqrt{0.75}$$

and

$$\sigma = \sqrt{0.75} = 0.866$$

INTERPRETATION In financial analysis, you can use the standard deviation to assess the degree of risk of an investment, as WORKED-OUT PROBLEM 4 illustrates.

WORKED-OUT PROBLEM 4 Suppose that you are deciding between two alternative investments. Investment A is a mutual fund whose portfolio consists of a combination of stocks that make up the Dow Jones Industrial Average. Investment B consists of shares of a growth stock. You estimate the returns (per $1,000 investment) for each investment alternative under three economic condition events (recession, stable economy, and expanding economy), and also provide your subjective probability of the occurrence of each economic condition as follows.

Estimated Return for Two Investments Under Three Economic Conditions

		Investment	
Probability	**Economic Event**	**Dow Jones Fund (A)**	**Growth Stock (B)**
0.2	Recession	–$100	–$200
0.5	Stable economy	+100	+50
0.3	Expanding economy	+250	+350

The mean or expected return for the two investments is computed as follows:

Mean = Sum of [Each value × the probability of each value]

Mean for the Dow Jones fund = (–100)(0.2) + (100)(0.5) + (250)(0.3) = $105

Mean for the growth stock = (–200)(0.2) + (50)(0.5) + (350)(0.3) = $90

You can calculate the standard deviation for the two investments, as shown in Tables 5.4 and 5.5.

TABLE 5.4

Computing the Variance and Standard Deviation for Dow Jones Fund (A)

Probability	Economic Event	Dow Jones Fund (A)	(Return – Mean Return)^2 × Probability
0.2	Recession	–\$100	$(-100 - 105)^2 \times (0.2) = (42{,}025) \times (0.2) = 8{,}405$
0.5	Stable economy	+100	$(100 - 105)^2 \times (0.5) = (25) \times (0.5) = 12.5$
0.3	Expanding economy	+250	$(250 - 105)^2 \times (0.3) = (21{,}025) \times (0.3) = 6{,}307.5$
			Total: (Variance) = 14,725

TABLE 5.5

Computing the Variance and Standard Deviation for Growth Stock (B)

Probability	Economic Event	Growth Stock (B)	(Return – Mean Return)^2 × Probability
0.2	Recession	–\$200	$(-200 - 90)^2 \times (0.2) = (84{,}100) \times (0.2) = 16{,}820$
0.5	Stable economy	+50	$(50 - 90)^2 \times (0.5) = (1{,}600) \times (0.5) = 800$
0.3	Expanding economy	+350	$(350 - 90)^2 \times (0.3)\ (67{,}600) \times (0.3) = 20{,}280$
			Total (Variance) = 37,900

σ = Square root of [Sum of (Squared differences between a value and the mean) × (Probability of the value)]

$$\sigma_A = \sqrt{(-100-105)^2(0.2)+(100-105)^2(0.5)+(250-105)^2(0.3)}$$
$$= \sqrt{14{,}725}$$
$$= \$121.35$$

$$\sigma_B = \sqrt{(-200-90)^2(0.2)+(50-90)^2(0.5)+(350-90)^2(0.3)}$$
$$= \sqrt{37{,}900}$$
$$= \$194.68$$

The Dow Jones fund has a higher mean return than the growth fund and also has a lower standard deviation, indicating less variation in the return under the different economic conditions. Having a higher mean return with less variation makes the Dow Jones fund a more desirable investment than the growth fund.

equation blackboard (optional)

To write the equations for the mean and standard deviation for a discrete probability distribution, you need the following symbols:

- An uppercase italic X, X, that represents a variable.
- An uppercase italic X with an italic lowercase i subscript, X_i, that represents the ith event associated with variable X.
- An uppercase italic N, N, that represents the number of elementary events for the variable X. (In Chapter 3, this symbol was called the population size.)

The symbol $P(X_i)$, which represents the probability of the event X_i.

The population mean, μ.

The population standard deviation σ.

Using these symbols creates these equations:

The mean of a probability distribution:

$$\mu = \sum_{i=1}^{N} X_i P(X_i)$$

The standard deviation of a probability distribution:

$$\sigma = \sqrt{\sum_{i=1}^{N} (X_i - \mu)^2 P(X_i)}$$

5.2 The Binomial and Poisson Probability Distributions

As mentioned in the previous section, probability distributions for certain types of discrete variables can be modeled using a mathematical formula. This section looks at two important discrete distributions that are widely used to compute probabilities. The first probability distribution, the binomial, is used for variables that have only two mutually exclusive events. The second probability distribution, the Poisson, is used when you are counting the number of outcomes that occur in a unit.

The Binomial Distribution

important point

CONCEPT The probability distribution for a discrete variable that meets these criteria:

- The variable is for a sample that consists of a fixed number of experimental trials (the sample size).
- The variable has only two mutually exclusive and collectively exhaustive events, typically labeled as success and failure.
- The probability of an event being classified as a success, p, and the probability of an event being classified as a failure, $1 - p$, are both constant in all experimental trials.
- The event (success or failure) of any single experimental trial is independent of (not influenced by) the event of any other trial.

EXAMPLE The coin tossing experiment described in WORKED-OUT PROBLEM 1 on page 80.

INTERPRETATION Using the binomial distribution avoids having to develop the probability distribution by using a table of outcomes and applying the multiplication rule, as was done in Section 4.3. This distribution also does not require that the probability of success is 0.5, thereby allowing you to use it in more situations than the method discussed in Section 5.1.

You typically determine binomial probabilities by either using the formula in the EQUATION BLACKBOARD on page 87, by using a table of binomial probabilities, or by using software functions that create customized tables (see the figure on page 89).

When the probability of success is 0.5, you can still use the table and multiplication rule method as was done in Table 5.1. Observe from the results of that table—that the probability of zero heads is 0.125, the probability of one head is 0.375, the probability of two heads is 0.375, and the probability of three heads is 0.125.

Binomial distributions can be symmetrical or skewed. Whenever $p = 0.5$, the binomial distribution will be symmetrical regardless of how large or small the value of the sample size, n. However, when $p \neq 0.5$, the distribution will be skewed. If $p < 0.5$, the distribution will be positive or right-skewed; if $p > 0.5$, the distribution will be negative or left-skewed. The distribution will become more symmetrical as p gets close to 0.5 and as the sample size, n, gets large.

The characteristics of the binomial distribution are:

Mean	The sample size (n) times the probability of success or $n \times p$, remembering that the sample size is the number of experimental trials.
Variance	The product of these three: sample size, probability of success, and probability of failure (1 – probability of success), or $n \times p \times (1 - p)$
Standard deviation	The square root of the variance, or $\sqrt{np(1-p)}$

equation blackboard (optional)

For the equation for the binomial distribution, use the symbols X (variable), n (sample size), and p (probability of success) previously introduced and add these symbols:

- A lowercase italic X, x, which represents the number of successes in the sample.
- The symbol $P(X = x \mid n, p)$, which represents the probability of the value x, given sample size n and probability of success p.

interested in math?

You use these symbols to form two separate expressions. One expression represents the number of ways you can get a certain number of successes in a certain number of trials:

$$\frac{n!}{x!(n-x)!}$$

(The symbol ! means factorial, where $n! = (n)(n-1)\ldots(1)$ so that 3! equals 6, $3 \times 2 \times 1$. 1! equals 1 and 0! is defined as being equal to 1.)

The second expression represents the probability of getting a certain number of successes in a certain number of trials *in a specific order*:

$$p^x \times (1-p)^{n-x}$$

Using these expressions forms the following equation:

$$P(X = x \mid n, p) = \frac{n!}{x!(n-x)!} p^x (1-p)^{n-x}$$

As an example, the calculations for determining the binomial probability of one head in three tosses of a fair coin (that is, for a problem in which $n = 3$, $p = 0.5$, and $x = 1$) are as follows:

$$P(X = 1) \mid n = 3, p = 0.5) = \frac{3!}{1!(3-1)!}(0.5)^1(1-0.5)^{3-1}$$

$$= \frac{3!}{1!(2)!}(0.5)^1(1-0.5)^2$$

$$= 3(0.5)(0.25) = 0.375$$

Using symbols previously introduced, you can write the equation for the mean and standard deviation of the binomial distribution:

$$\mu = np$$

and

$$\sigma = \sqrt{np(1-p)}$$

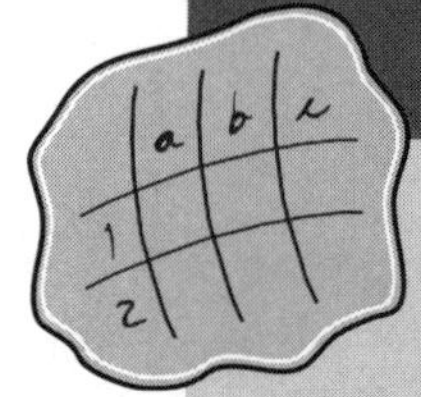

spreadsheet solution

Binomial Probabilities

Chapter 5 Binomial contains the spreadsheet shown on page 89 that calculates binomial probabilities. Experiment with this chart by changing the sample size in cell B4 and the probability of success in cell B5.

Best Practices

Use the **BINOM.DIST(*number of successes, sample size, probability of success, cumulative?*)** function to calculate binomial probabilities.

How-Tos

In the Binomial Probabilities Table, *cumulative?* is set to **False** in the column B BINOM.DIST functions to calculate the probability for exactly the *number of successes*.

WORKED-OUT PROBLEM 5 An online social networking website defines success if a Web surfer stays and views its website for more than three minutes. Suppose that the probability that the surfer does stay for more than three minutes is 0.16. What is the probability that at least four (either four or five) of the next five surfers will stay for more than three minutes?

You need to sum the probabilities of four surfers staying and five surfers staying in order to determine the probabilities that at least four surfers stay.

These sums can be found in the $X = 4$ and $X = 5$ rows in the Binomial Probabillities Table of the Chapter 5 Binomial spreadsheet.

	A	B
1	Binomial Probabilities	
2		
3	Data	
4	Sample size	5
5	Probability of success	0.16
6		
7	Statistics	
8	Mean	0.8
9	Variance	0.672
10	Standard deviation	0.8198
11		
12	Binomial Probabilities Table	
13	X	P(X)
14	0	0.4182
15	1	0.3983
16	2	0.1517
17	3	0.0289
18	4	0.0028
19	5	0.0001

From the spreadsheet table:

$P(X = 4|n = 5, p = 0.16) = 0.0028$

$P(X = 5|n = 5, p = 0.16) = 0.0001$

Therefore, the probability of four or more surfers staying and viewing the social networking website for more than three minutes is 0.0029 (which you compute by adding 0.0028 and 0.0001) or 0.29%.

The Poisson Distribution

CONCEPT The probability distribution for a discrete variable that meets these criteria:

- You are counting the number of times a particular event occurs in a unit.
- The probability that an event occurs in a particular unit is the same for all other units.
- The number of events that occur in a unit is independent of the number of events that occur in other units.
- As the unit gets smaller, the probability that two or more events will occur in that unit approaches zero.

EXAMPLES Number of computer network failures per day, number of surface defects per square yard of floor covering, the number of customers arriving at a bank during the 12 noon to 1 p.m. hour, the number of fleas on the body of a dog.

INTERPRETATION To use the Poisson distribution, you define an area of opportunity, a continuous unit of area, time, or volume in which more than one event can occur. The Poisson distribution can model many variables that count the number of defects per area of opportunity or count the number of times items are processed from a waiting line.

You determine Poisson probabilities by applying the formula in the EQUATION BLACKBOARD on page 91, by using a table of Poisson values, or by using software functions that create customized tables (see the figure below). The characteristics of the Poisson distribution are:

Mean	The population mean, λ.
Variance	In the Poisson distribution, the variance is equal to the population mean, λ.
Standard deviation	The square root of the variance, or $\sqrt{\lambda}$.

WORKED-OUT PROBLEM 6 You want to determine the probabilities that a specific number of customers will arrive at a bank branch in a one-minute interval during the lunch hour: Will zero customers arrive, one customer, two customers, and so on? You determine that you can use the Poisson distribution because of the following reasons:

- The variable is a count per unit, customers per minute.
- You assume that the probability that a customer arrives during a specific one-minute interval is the same as the probability for all the other one-minute intervals.
- Each customer's arrival has no effect on (is independent of) all other arrivals.
- The probability that two or more customers will arrive in a given time period approaches zero as the time interval decreases from one minute.

Using historical data, you determine that the mean number of arrivals of customers is three per minute during the lunch hour. You use a spreadsheet table (shown below) to calculate the Poisson probabilities, using 3 as the value for mean (expected) number of successes:

	A	B
1	Poisson Probabilities	
2		
3	Data	
4	Mean (expected) number of successes	3
5		
6	Poisson Probabilities Table	
7	X	P(X)
8	0	0.0498
9	1	0.1494
10	2	0.2240
11	3	0.2240
12	4	0.1680
13	5	0.1008
14	6	0.0504
15	7	0.0216
16	8	0.0081
17	9	0.0027
18	10	0.0008
19	11	0.0002
20	12	0.0001
21	13	0.0000
22	14	0.0000
23	15	0.0000

From the results, you observe the following:

- The probability of zero arrivals is 0.0498.
- The probability of one arrival is 0.1494.
- The probability of two arrivals is 0.2240.

Therefore, the probability of two or fewer customer arrivals per minute at the bank during the lunch hour is 0.4232, the sum of the probabilities for zero, one, and two arrivals (0.0498 + 0.1494 + 0.2240 = 0.4232).

equation blackboard (optional)

interested in math?

For the equation for the Poisson distribution, use the symbols X (variable), n (sample size), p (probability of success) previously introduced and add these symbols:

- A lowercase italic E, e, which represents the mathematical constant approximated by the value 2.71828.
- A lowercase Greek symbol lambda, λ, which represents the mean number of times that the event occurs per area of opportunity.
- A lowercase italic X, x, which represents the number of times the event occurs per area of opportunity.
- The symbol $P(X = x \mid \lambda)$, which represents the probability of x, given λ.

Using these symbols forms the following equation:

$$P(X = x \mid \lambda) = \frac{e^{-\lambda} \lambda^{x}}{x!}$$

As an example, the calculations for determining the Poisson probability of exactly two arrivals in the next minute given a mean of three arrivals per minute is as follows:

$$\begin{aligned} P(X = 2 \mid \lambda = 3) &= \frac{e^{-3}(3)^2}{2!} \\ &= \frac{(2.71828)^{-3}(3)^2}{2!} \\ &= \frac{(0.049787)(9)}{(2)} \\ &= 0.2240 \end{aligned}$$

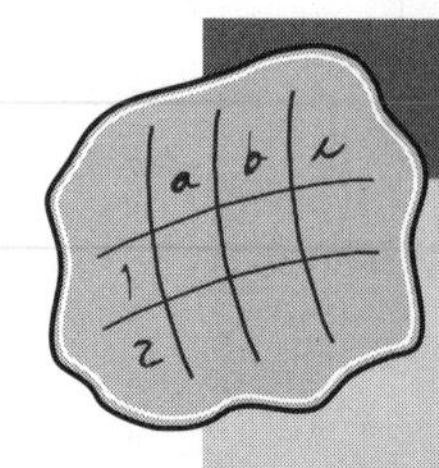

spreadsheet solution

Poisson Probabilities

Chapter 5 Poisson contains the spreadsheet shown on page 90 that calculates Poisson probabilities. Experiment with this chart by changing the mean (expected) number of success in cell B4.

Best Practices

Use the POISSON.DIST(***number of successes, mean number of successes, cumulative?***) function to calculate Poisson probabilities.

How Tos

In the Poisson Probabilities Table, *cumulative?* is set to **False** in the column B POISSON.DIST functions to calculate the probability for exactly the *number of successes*.

5.3 Continuous Probability Distributions and the Normal Distribution

Probability distributions can also be developed to model continuous variables. The exact mathematical expression for the probability distribution for a continuous variable involves integral calculus and is not shown in this book.

Probability Distribution for a Continuous Variable

CONCEPT The area under a curve that represents the probabilities for a continuous variable.

EXAMPLE See the example for the normal distribution on page 93.

INTERPRETATION Probability distributions for a continuous variable differ from discrete distributions in several important ways:

important point

- An event can take on any value within the range of the variable and not just an integer value.
- The probability of any specific value is zero.
- Probabilities are expressed in terms of an area under a curve that represents the continuous distribution.

One continuous distribution, the **normal distribution**, is especially important in statistics because it can model many different continuous variables.

Normal Distribution

CONCEPT The probability distribution for a continuous variable that meets these criteria:

- The graphed curve of the distribution is bell-shaped and symmetrical.
- The mean, median, and mode are all the same value.
- The population mean, μ, and the population standard deviation, σ, determine probabilities.
- The distribution extends from negative to positive infinity. (The distribution has an infinite range.)
- Probabilities are always cumulative and expressed as inequalities, such as $P < X$ or $P \geq X$, where X is a value for the variable.

EXAMPLE The normal distribution appears as a bell-shaped curve as shown in the following figure.

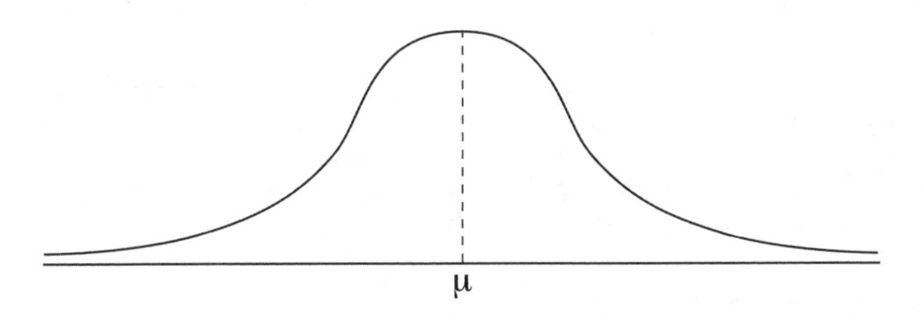

INTERPRETATION The importance of the normal distribution to statistics cannot be overstated. Probabilities associated with variables as diverse as physical characteristics such as height and weight, scores on standardized exams, and the dimension of industrial parts, tend to follow a normal distribution. Under certain circumstances, the normal distribution also approximates various discrete probability distributions such as the binomial and Poisson distributions. In addition, the normal distribution provides the basis for classical statistical inference discussed in Chapters 6 through 9.

You determine normal probabilities by using a table of normal probabilities (such as Table C.1 in Appendix C) or by using software functions. Normal probability tables such as Table C.1 and some software functions use a standardized normal distribution that requires you to convert an X value of a variable to its corresponding Z score (see Section 3.3). You perform this conversion by subtracting the population mean μ from the X value and dividing the resulting

difference by the population standard deviation σ, expressed algebraically as follows:

$$Z = \frac{X - \mu}{\sigma}$$

When the mean is 0 and the standard deviation is 1, the X value and Z score will be the same and no conversion is necessary.

WORKED-OUT PROBLEM 7 Packages of chocolate candies have a labeled weight of 6 ounces. In order to ensure that very few packages have a weight below 6 ounces, the filling process provides a mean weight above 6 ounces. In the past, the mean weight has been 6.15 ounces with a standard deviation of 0.05 ounce. Suppose you want to determine the probability that a single package of chocolate candies will weigh between 6.15 and 6.20 ounces. To determine this probability, you use Table C.1, the table of the probabilities of the cumulative standardized normal distribution.

To use Table C.1, you must first convert the weights to Z scores by subtracting the mean and then dividing by the standard deviation, as shown here:

$$Z(\text{lower}) = \frac{6.15 - 6.15}{0.05} = 0 \qquad Z(\text{upper}) = \frac{6.20 - 6.15}{0.05} = 1.0$$

Therefore, you need to determine the probability that corresponds to the area between 0 and +1 Z units (standard deviations). To do this, you take the cumulative probability associated with 0 Z units and subtract it from the cumulative probability associated with +1 Z units. Using Table C.1, you determine that these probabilities are 0.8413 and 0.5000, respectively (see the following table).

Finding a Cumulative Area Under the Normal Curve

	Cumulative Probabilities									
Z	**.00**	**.01**	**.02**	**.03**	**.04**	**.05**	**.06**	**.07**	**.08**	**.09**
0.0	.5000	.5040	.5080	.5120	.5160	.5199	.5239	.5279	.5319	.5359
0.1	.5398	.5438	.5478	.5517	.5557	.5596	.5636	.5675	.5714	.5753
0.2	.5793	.5832	.5871	.5910	.5948	.5987	.6026	.6064	.6103	.6141
0.3	.6179	.6217	.6255	.6293	.6331	.6368	.6406	.6443	.6480	.6517
0.4	.6554	.6591	.6628	.6664	.6700	.6736	.6772	.6808	.6844	.6879
0.5	.6915	.6950	.6985	.7019	.7054	.7088	.7123	.7157	.7190	.7224
0.6	.7257	.7291	.7324	.7357	.7389	.7422	.7454	.7486	.7518	.7549
0.7	.7580	.7612	.7642	.7673	.7704	.7734	.7764	.7794	.7823	.7852
0.8	.7881	.7910	.7939	.7967	.7995	.8023	.8051	.8078	.8106	.8133
0.9	.8159	.8186	.8212	.8238	.8264	.8289	.8315	.8340	.8365	.8389
1.0	.8413	.8438	.8461	.8485	.8508	.8531	.8554	.8577	.8599	.8621

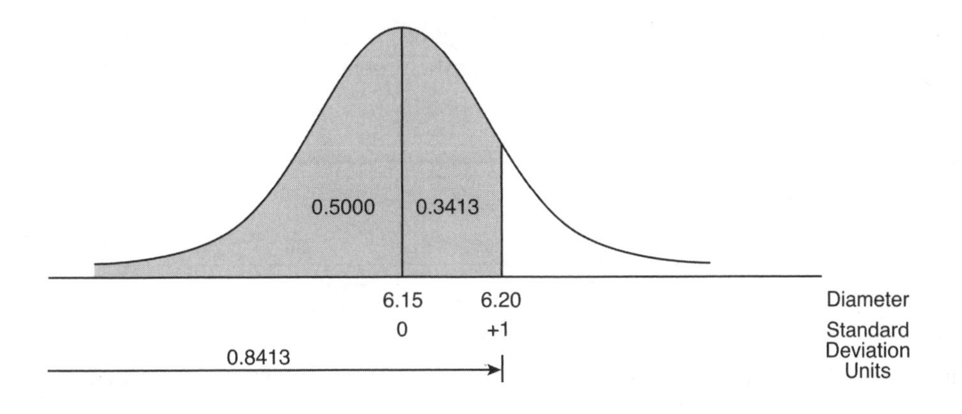

Therefore, the probability that a single package of chocolate candies will weigh between 6.15 and 6.20 ounces is 0.3413 (0.8413 – 0.5000 = 0.3413). Using the cumulative probability associated with +1 Z units, you can also calculate that the probability that a package weighs more than 6.20 ounces is 0.1587 (1 – 0.8413). (The **Chapter 5 Normal** worksheet discussed on page 99 can also be used to produce these results.)

Using Standard Deviation Units

Because of the equivalence between Z scores and standard deviation units, probabilities of the normal distribution are often expressed as ranges of plus-or-minus standard deviation units. Such probabilities can be determined directly from Table C.1, the table of the probabilities of the cumulative standardized normal distribution.

For example, to determine the normal probability associated with the range ±3 standard deviations, you would use Table C.1 to look up the probabilities associated with Z = –3.00 and Z = +3.00:

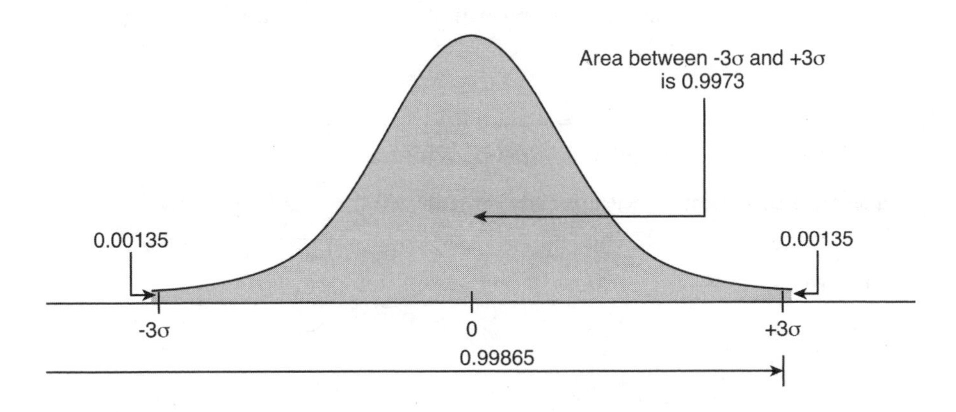

Table 5.6 represents the appropriate portion of Table C.1 for Z = –3.00. From this table excerpt, you can determine that the probability of a value less than Z = –3 units is 0.00135.

TABLE 5.6

Partial Table C.1 for Obtaining a Cumulative Area Below –3 Z Units

Z	.00	.01	.02	.03	.04	.05	.06	.07	.08	.09
.	.	.	.	.	.	.	.	.	.	.
.	.	.	.	.	.	.	.	.	.	.
.	.	.	.	.	.	.	.	.	.	.
–3.0 →	0.00135	0.00131	0.00126	0.00122	0.00118	0.00114	0.00111	0.00107	0.00103	0.00100

Source: Extracted from Table C.1

Table 5.7 represents the appropriate portion of Table C.1 for Z = +3.00. From this table excerpt, you can determine that the probability of a value less than Z = +3 units is 0.99865.

TABLE 5.7

Partial Table C.1 for Obtaining a Cumulative Area Below +3 Z Units

Z	.00	.01	.02	.03	.04	.05	.06	.07	.08	.09
.	.	.	.	.	.	.	.	.	.	.
.	.	.	.	.	.	.	.	.	.	.
.	.	.	.	.	.	.	.	.	.	.
+3.0 →	0.99865	0.99869	0.99874	0.99878	0.99882	0.99886	0.99889	0.99893	0.99897	0.99900

Source: Extracted from Table C.1

Therefore, the probability associated with the range plus-or-minus three standard deviations in a normal distribution is 0.9973 (0.99865 – 0.00135). Stated another way, the probability is 0.0027 (2.7 out of a thousand chance) that a value will not be within the range of plus-or-minus three standard deviations. Table 5.8 summarizes probabilities for several different ranges of standard deviation units.

TABLE 5.8

Probabilities for Different Standard Deviation Ranges

Standard Deviation Unit Ranges	Probability or Area Outside These Units	Probability or Area Within These Units
-1σ to $+1\sigma$	0.3174	0.6826
-2σ to $+2\sigma$	0.0455	0.9545
-3σ to $+3\sigma$	0.0027	0.9973
-6σ to $+6\sigma$	0.000000002	0.999999998

Finding the *Z* Value from the Area Under the Normal Curve

Each of the previous examples involved using the normal distribution table to find an area under the normal curve that corresponded to a specific *Z* value. In many circumstances, you want to do the opposite of this and find the *Z* value that corresponds to a specific area. For example, you can find the *Z* value that corresponds to a cumulative area of 1%, 5%, 95%, or 99%. You can also find lower and upper *Z* values between which 95% of the area under the curve is contained.

To find the *Z* value that corresponds to a cumulative area, you locate the cumulative area in the body of the normal table, or the closest value to the cumulative area you want to find, and then determine the *Z* value that corresponds to this cumulative area.

WORKED-OUT PROBLEM 8 You want to find the *Z* values such that 95% of the normal curve is contained between a lower *Z* value and an upper *Z* value with 2.5% below the lower *Z* value, and 2.5% above the upper *Z* value. Using the following figure, you determine that you need to find the *Z* value that corresponds to a cumulative area of 0.025 and the *Z* value that corresponds to a cumulative area of 0.975.

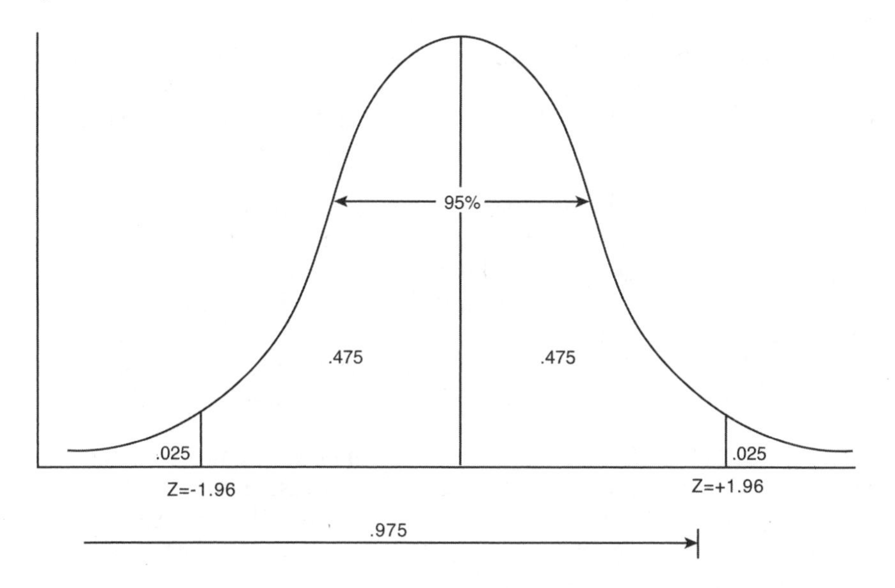

Table 5.9 contains a portion of Table C.1 that is needed to find the *Z* value that corresponds to a cumulative area of 0.025. Table 5.10 contains a portion of Table C.1 that is needed to find the *Z* value that corresponds to a cumulative area of 0.975.

TABLE 5.9

Partial Table C.1 for Finding Z Value That Corresponds to a Cumulative Area of 0.025

Z	.00	.01	.02	.03	.04	.05	.06	.07	.08	.09
.	.	.	.	.	.	.	.	.	.	.
.	.	.	.	.	.	.	.	.	.	.
.	.	.	.	.	.	.	.	.	.	.
−2.0	0.0228	0.0222	0.0217	0.0212	0.0207	0.0202	0.0197	0.0192	0.0188	0.0183
−1.9	0.0287	0.0281	0.0274	0.0268	0.0262	0.0256	0.0250	0.0244	0.0239	0.0233

TABLE 5.10

Partial Table C.1 for Finding Z Value That Corresponds to a Cumulative Area of 0.975

Z	.00	.01	.02	.03	.04	.05	.06	.07	.08	.09
.	.	.	.	.	.	.	.	.	.	.
.	.	.	.	.	.	.	.	.	.	.
.	.	.	.	.	.	.	.	.	.	.
1.9	0.9713	0.9719	0.9726	0.9732	0.9738	0.9744	0.9750	0.9756	0.9761	0.9767
2.0	0.9772	0.9778	0.9783	0.9788	0.9793	0.9798	0.9803	0.9808	0.9812	0.9817

To find the Z value that corresponds to a cumulative area of 0.025, you look in the body of Table 5.9 until you see the value of 0.025. Then you determine the row and column that this value corresponds to. Locating the value of 0.025, you see that it is located in the –1.9 row and the .06 column. Thus, the Z value that corresponds to a cumulative area of 0.025 is –1.96.

To find the Z value that corresponds to a cumulative area of 0.975, you look in the body of Table 5.10 until you see the value of 0.975. Then you determine the corresponding row and column that this value belongs to. Locating the value of 0.975, you see that it is in the 1.9 row and the .06 column. Thus the Z value that corresponds to a cumulative area of 0.975 is 1.96. Taking this result along with the Z value of –1.96 for a cumulative area of 0.025 means that 95% of all the values will be between $Z = -1.96$ and $Z = 1.96$.

WORKED-OUT PROBLEM 9 You want to find the weights that will contain 95% of the packages of chocolate candy first discussed in WORKED-OUT PROBLEM 7 on page 94. In order to do so, you need to determine X in the formula

$$Z = \frac{X - \mu}{\sigma}$$

Solving this formula for X, you have

$$X = \mu + Z\sigma$$

Because the mean weight is 6.15 ounces and the standard deviation is 0.05 ounce, and 95% of the packages will be contained between −1.96 and +1.96 standard deviation (Z) units, the interval that will contain 95% of the packages will be between

$$6.15 + (-1.96)(0.05) = 6.15 - 0.098 = 6.052 \text{ ounces}$$

and

$$6.15 + (+1.96)(0.05) = 6.15 + 0.098 = 6.248 \text{ ounces}$$

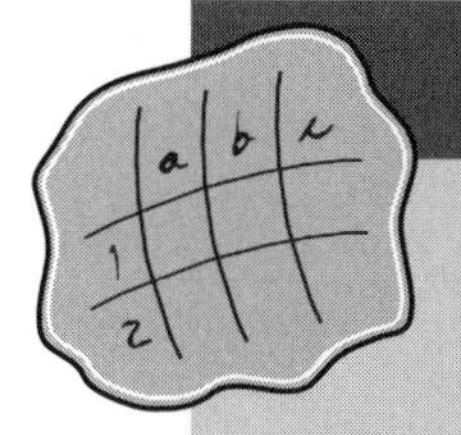

spreadsheet solution

Normal Probabilities

Chapter 5 Normal contains the spreadsheet shown below that calculates various normal probabilities. Experiment with this chart by changing the mean in cell B4 and the standard deviation in cell B5.

Best Practices

Use the **STANDARDIZE**, **NORM.DIST**, **NORM.S.INV**, and **NORM.INV** functions to calculate values associated with normal probabilities.

How-Tos

Tip FT2 in Appendix D explains how to enter the functions used in **Chapter 5 Normal**.

Tip ADV3 in Appendix E explains how the spreadsheet uses the ampersand operator (&) to produce some of the labels in columns A and D.

	A	B	C	D	E
1	Normal Probabilities				
2					
3	Common Data			Probability for a Range	
4	Mean	6.15		From X Value	6.15
5	Standard Deviation	0.05		To X Value	6.2
6				Z Value for 6.15	0
7	Probability for X <=			Z Value for 6.2	1
8	X Value	6.2		P(X<=6.15)	0.5000
9	Z Value	1		P(X<=6.2)	0.8413
10	P(X<=6.2)	0.8413		P(6.15<=X<=6.2)	0.3413
11					
12	Probability for X >			Find Z and X Given a Cumulative Pctage.	
13	X Value	6.2		Cumulative Percentage	2.50%
14	Z Value	1		Z Value	-1.96
15	P(X>6.2)	0.1587		X Value	6.05
16					
17				Find Z and X Values Given a Percentage	
18				Percentage	95.00%
19				Z Value	-1.96
20				Lower X Value	6.05
21				Upper X Value	6.25

5.4 The Normal Probability Plot

In order to use many inferential statistical methods, you must determine if a set of data approximately follows a normal distribution. One technique for making this determination is the **normal probability plot**.

CONCEPT A graph that plots the relationship between ranked data values and the Z scores that these values would correspond to if the set of data values follows a normal distribution. If the data values follow a normal distribution, the graph will be linear (a straight line), as shown in the following example.

EXAMPLES

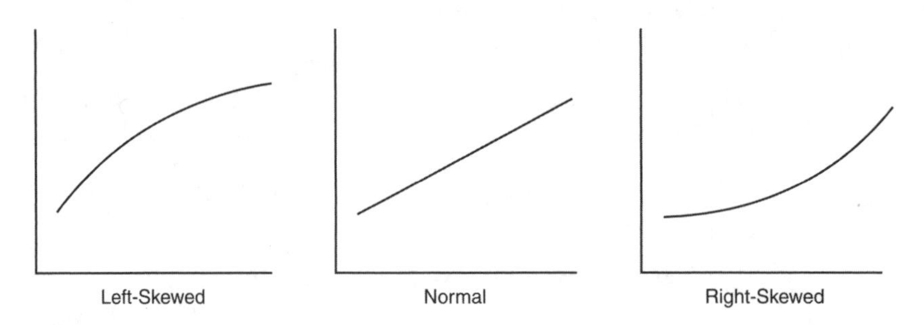

INTERPRETATION Normal probability plots are based on the idea that the Z scores for the ranked values increase at a predictable rate for data that follow a normal distribution. The exact details to produce a normal probability plot can vary, but one common approach is called the **quantile–quantile plot**. In this method, each value (ranked from lowest to highest) is plotted on the vertical Y axis and its transformed Z score is plotted on the horizontal X axis. If the data are normally distributed, a plot of the data ranked from lowest to highest will follow a straight line. As shown in the preceding figure, if the data are left-skewed, the curve will rise more rapidly at first, and then level off. If the data are right-skewed, the data will rise more slowly at first, and then rise at a faster rate for higher values of the variable being plotted.

Restaurants

WORKED-OUT PROBLEM 10 You seek to determine whether the city restaurant meal cost data first used in WORKED-OUT PROBLEM 7 in Chapter 3 (see page 43) follows a normal distribution. You use Microsoft Excel to produce the following normal probability plot:

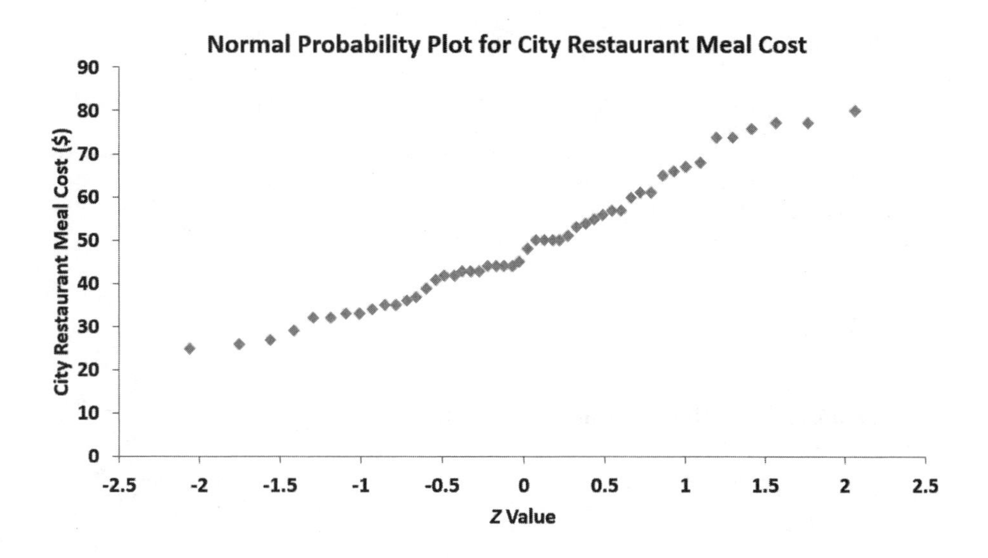

Consistent with the results of the histogram in Section 2.2, the approximate straight line that the data follow in this normal probability plot appears to indicate that the cost of a meal at city restaurants is approximately normally distributed.

Important Equations

The mean of a probability distribution:

(5.1) $\mu = \sum_{i=1}^{N} X_i P(X_i)$

The standard deviation of a discrete probability distribution:

(5.2) $\sigma = \sqrt{\sum_{i=1}^{N} (X_i - \mu)^2 P(X_i)}$

The binomial distribution:

(5.3) $P(X = x) | n, p) = \frac{n!}{x!\,(n-x)!} p^x (1-p)^{n-x}$

The mean of the binomial distribution:

(5.4) $\mu = np$

The standard deviation of the binomial distribution:

(5.5) $\sigma_x = \sqrt{np(1-p)}$

The Poisson distribution:

(5.6) $P(X = x \mid \lambda) = \frac{e^{-\lambda}\lambda^x}{x!}$

The normal distribution: finding a *Z* value

(5.7) $Z = \frac{X - \mu}{\sigma}$

The normal distribution: finding an *X* value

(5.8) $X = \mu + Z\sigma$

One-Minute Summary

Probability Distributions

Discrete probability distributions

- Expected value
- Variance σ^2 and standard deviation σ

Is there a fixed sample size *n* and is each observation classified into one of two categories?

- If yes, use the binomial distribution, subject to other conditions.
- If no, use the Poisson distribution, subject to other conditions.

Continuous probability distributions

- Normal distribution
- Normal probability plot

Test Yourself

Short Answers

1. The sum of the probabilities of all the events in a probability distribution is equal to:
   (a) 0
   (b) the mean
   (c) the standard deviation
   (d) 1

2. The largest number of possible successes in a binomial distribution is:
   (a) 0
   (b) 1
   (c) *n*
   (d) infinite

3. The smallest number of possible successes in a binomial distribution is:
   (a) 0
   (b) 1
   (c) *n*
   (d) infinite

4. Which of the following about the binomial distribution is not a true statement?
   (a) The probability of success must be constant from trial to trial.
   (b) Each outcome is independent of the other.
   (c) Each outcome may be classified as either "success" or "failure."
   (d) The variable of interest is continuous.

5. Whenever $p = 0.5$, the binomial distribution will:
   (a) always be symmetric
   (b) be symmetric only if *n* is large
   (c) be right-skewed
   (d) be left-skewed

6. What type of probability distribution will the consulting firm most likely employ to analyze the insurance claims in the following problem?

   An insurance company has called a consulting firm to determine whether the company has an unusually high number of false insurance claims. It is known that the industry proportion for false claims is 6%.

The consulting firm has decided to randomly and independently sample 50 of the company's insurance claims. They believe that the number of false claims from the sample will yield the information the company desires.

(a) Binomial distribution

(b) Poisson distribution

(c) Normal distribution

(d) None of the above

7. What type of probability distribution will most likely be used to analyze warranty repair needs on new cars in the following problem?

   The service manager for a new automobile dealership reviewed dealership records of the past 20 sales of new cars to determine the number of warranty repairs he will be called on to perform in the next 30 days. Corporate reports indicate that the probability any one of their new cars needs a warranty repair in the first 30 days is 0.035. The manager assumes that calls for warranty repair are independent of one another and is interested in predicting the number of warranty repairs he will be called on to perform in the next 30 days for this batch of 20 new cars sold.

   (a) Binomial distribution

   (b) Poisson distribution

   (c) Normal distribution

   (d) None of the above

8. The quality control manager of Marilyn's Cookies is inspecting a batch of chocolate chip cookies. When the production process is in control, the mean number of chocolate chip parts per cookie is 6.0. The manager is interested in analyzing the probability that any particular cookie being inspected has fewer than 10.0 chip parts. What probability distribution should be used?

   (a) Binomial distribution

   (b) Poisson distribution

   (c) Normal distribution

   (d) None of the above

9. The smallest number of possible successes in a Poisson distribution is:

   (a) 0

   (b) 1

   (c) n

   (d) infinite

10. Based on past experience, the time you spend on emails per day has a mean of 30 minutes and a standard deviation of 10 minutes. To compute the probability of spending at least 12 minutes on emails, you can use what probability distribution?
    (a) Binomial distribution
    (b) Poisson distribution
    (c) Normal distribution
    (d) None of the above

11. A computer lab at a university has ten personal computers. Based on past experience, the probability that any one of them will require repair on a given day is 0.05. To find the probability that exactly two of the computers will require repair on a given day, you can use what probability distribution?
    (a) Binomial distribution
    (b) Poisson distribution
    (c) Normal distribution
    (d) None of the above

12. The mean number of customers who arrive per minute at any one of the checkout counters of a grocery store is 1.8. What probability distribution can be used to find out the probability that there will be no customers arriving at a checkout counter in the next minute?
    (a) Binomial distribution
    (b) Poisson distribution
    (c) Normal distribution
    (d) None of the above

13. A multiple-choice test has 25 questions. There are four choices for each question. A student who has not studied for the test decides to answer all questions by randomly choosing one of the four choices. What probability distribution can be used to compute his chance of correctly answering at least 15 questions?
    (a) Binomial distribution
    (b) Poisson distribution
    (c) Normal distribution
    (d) None of the above

14. Which of the following about the normal distribution are true?
    (a) Theoretically, the mean, median, and mode are the same.
    (b) About 99.7% of the values fall within three standard deviations from the mean.
    (c) It is defined by two characteristics μ and σ.
    (d) All of the above are true.

15. Which of the following about the normal distribution is not true?
    (a) Theoretically, the mean, median, and mode are the same.
    (b) About two-thirds of the observations fall within one standard deviation from the mean.
    (c) It is a discrete probability distribution.
    (d) Its parameters are the mean, μ, and standard deviation, σ.

16. The probability that Z is less than –1.0 is ________ the probability that Z is greater than +1.0.
    (a) less than
    (b) the same as
    (c) greater than

17. The normal distribution is ________ in shape:
    (a) right-skewed
    (b) left-skewed
    (c) symmetric

18. If a particular set of data is approximately normally distributed, you would find that approximately:
    (a) 2 of every 3 observations would fall between 1 standard deviation around the mean
    (b) 4 of every 5 observations would fall between 1.28 standard deviations around the mean
    (c) 19 of every 20 observations would fall between 2 standard deviations around the mean
    (d) All of the above

19. Given that X is a normally distributed variable with a mean of 50 and a standard deviation of 2, the probability that X is between 47 and 54 is ________.

Answer True or False:

20. Theoretically, the mean, median, and the mode are all equal for a normal distribution.

21. Another name for the mean of a probability distribution is its expected value.

22. The diameters of 100 randomly selected bolts follow a binomial distribution.

23. If the data values are normally distributed, the normal probability plot will follow a straight line.

Answers to Test Yourself Short Answers

1. d
2. c
3. a
4. d
5. a
6. a
7. a
8. b
9. a
10. c
11. a
12. b
13. a
14. d
15. c
16. b
17. c
18. d
19. 0.9104
20. True
21. True
22. False
23. True

Problems

1. Given the following probability distributions:

Distribution A		Distribution B	
X	*P(X)*	*X*	*P(X)*
0	0.20	0	0.10
1	0.20	1	0.20
2	0.20	2	0.40
3	0.20	3	0.20
4	0.20	4	0.10

   (a) Compute the expected value of each distribution.
   (b) Compute the standard deviation of each distribution.
   (c) Compare the results of distributions *A* and *B*.

2. In the carnival game Under-or-Over-Seven, a pair of fair dice is rolled once, and the resulting sum determines whether the player wins or loses his or her bet. For example, the player can bet $1 that the sum will be under 7—that is, 2, 3, 4, 5, or 6. For this bet, the player wins $1 if the result is under 7 and loses $1 if the outcome equals or is greater than 7. Similarly, the player can bet $1 that the sum will be over 7—

that is, 8, 9, 10, 11, or 12. Here, the player wins $1 if the result is over 7 but loses $1 if the result is 7 or under. A third method of play is to bet $1 on the outcome 7. For this bet, the player wins $4 if the result of the roll is 7 and loses $1 otherwise.

(a) Construct the probability distribution representing the different outcomes that are possible for a $1 bet on being under 7.

(b) Construct the probability distribution representing the different outcomes that are possible for a $1 bet on being over 7.

(c) Construct the probability distribution representing the different outcomes that are possible for a $1 bet on 7.

(d) Show that the expected long-run profit (or loss) to the player is the same, no matter which method of play is used.

3. The number of arrivals per minute at a bank located in the central business district of a large city was recorded over a period of 200 minutes with the following results:

Arrivals	Frequency
0	14
1	31
2	47
3	41
4	29
5	21
6	10
7	5
8	2

(a) Compute the expected number of arrivals per day.

(b) Compute the standard deviation.

4. Suppose that a judge's decisions are upheld by an appeals court 90% of the time. In her next ten decisions, what is the probability that

(a) eight of her decisions are upheld by an appeals court?

(b) all ten of her decisions are upheld by an appeals court?

(c) eight or more of her decisions are upheld by an appeals court?

5. A venture capitalist firm that specializes in funding risky high-technology startup companies has determined that only one in ten of its companies is a "success" that makes a substantive profit within

six years. Given this historical record, what is the probability that in the next three startups it finances:

(a) The firm will have exactly one success?

(b) Exactly two successes?

(c) Less than two successes?

(d) At least two successes?

6. Accuracy in taking orders at a drive-through window is important for fast-food chains. Periodically, *QSR Magazine* publishes the results of a survey that measures accuracy, defined as the percentage of orders that are filled correctly. In a recent month, the percentage of orders filled correctly at Burger King was approximately 82.3%. (Source: **qsrmagazine.com/content/drive-thru-performance-study-order-accuracy**.)

   Suppose that you go to the drive-through window at Burger King and place an order. Two friends of yours independently place orders at the drive-through window at the same Burger King. What are the probabilities that:

   (a) All three of the three orders will be filled correctly?

   (b) None of the three orders will be filled correctly?

   (c) At least two of the three orders will be filled correctly?

   (d) What are the mean and standard deviation of the binomial distribution for the number of orders filled correctly?

7. The number of power outages at a power plant has a Poisson distribution with a mean of four outages per year. What is the probability that in a year there will be

   (a) no power outages?

   (b) four power outages?

   (c) at least three power outages?

8. The quality control manager of Marilyn's Cookies is inspecting a batch of chocolate-chip cookies that has just been baked. If the production process is in control, the mean number of chip parts per cookie is 6.0. What is the probability that in any particular cookie being inspected there are

   (a) less than five chip parts?

   (b) exactly five chip parts?

   (c) five or more chip parts?

   (d) either four or five chip parts?

9. The U.S. Department of Transportation maintains statistics for mishandled bags per 1,000 airline passengers. In a recent month, airlines had 2.77 mishandled bags per 1,000 passengers. What is the probability that in the next 1,000 passengers, airlines will have
   (a) no mishandled bags?
   (b) at least one mishandled bag?
   (c) at least two mishandled bags?

10. Given that X is a normally distributed variable with a mean of 50 and a standard deviation of 2, what is the probability that
    (a) X is between 47 and 54?
    (b) X is less than 55?
    (c) There is a 90% chance that X will be less than what value?

11. A set of final examination grades in an introductory statistics course is normally distributed, with a mean of 73 and a standard deviation of 8.
    (a) What is the probability of getting a grade below 91 on this exam?
    (b) What is the probability that a student scored between 65 and 89?
    (c) The probability is 5% that a student taking the test scores higher than what grade?
    (d) If the professor grades on a curve (that is, gives A's to the top 10% of the class, regardless of the score), are you better off with a grade of 81 on this exam or a grade of 68 on a different exam, where the mean is 62 and the standard deviation is 3? Explain.

12. The owner of a fish market determined that the mean weight for salmon is 12.3 pounds with a standard deviation of 2 pounds. Assuming the weights of salmon are normally distributed, what is the probability that
    (a) a randomly selected salmon will weigh between 12 and 15 pounds?
    (b) a randomly selected salmon will weigh less than 10 pounds?
    (c) 95% of the salmon will weigh between what two values?

Restaurants

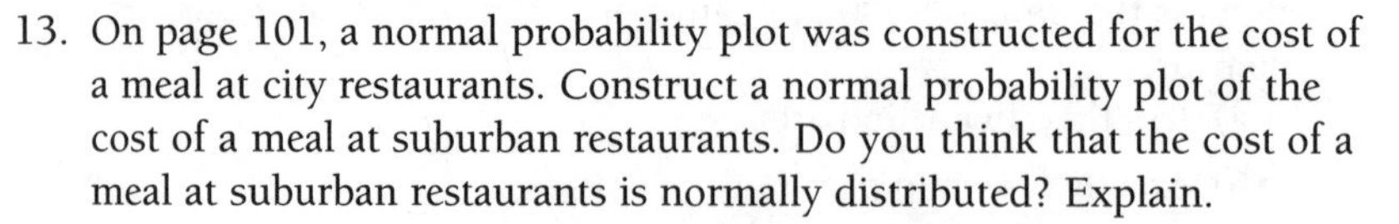

13. On page 101, a normal probability plot was constructed for the cost of a meal at city restaurants. Construct a normal probability plot of the cost of a meal at suburban restaurants. Do you think that the cost of a meal at suburban restaurants is normally distributed? Explain.

NBACost

14. Is the cost of attending an NBA basketball game normally distributed? Construct a normal probability plot of the cost of attending an NBA basketball game. Do you think that the cost of attending an NBA basketball game is normally distributed? Explain.

15. The file **DomesticBeer** contains the percentage alcohol, number of calories per 12 ounces, and number of carbohydrates (in grams) per 12 ounces for 152 of the best-selling domestic beers in the United States. (Data extracted from **www.beer100.com/beercalories.htm**, 20 March 2013.) Do you think any of these variables are normally distributed? Explain.

Answers to Test Yourself Problems

1. (a) A: $\mu = 2$, B: $\mu = 2$.

   (b) A: $\sigma = 1.414$, B: $\sigma = 1.095$.

   (c) Distribution A is uniform and symmetric; Distribution B is symmetric and has a smaller standard deviation than distribution A.

2. (a)

X	$P(X)$
\$ – 1	21/36
\$ + 1	15/36

   (b)

X	$P(X)$
\$ – 1	21/36
\$ + 1	15/36

   (c)

X	$P(X)$
\$ – 1	30/36
\$ + 4	6/36

   (d) \$–0.167 for each method of play.

3. (a) 2.90 (b) 1.772
4. (a) 0.1937 (b) 0.3487 (c).0.9298
5. (a) 0.243 (b) 0.027 (c) 0.972 (d) 0.028
6. (a) 0.5574 (b) 0.0055 (c) 0.9171 (d) 2.469 and 0.6611
7. (a) 0.0183 (b) 0.1954 (c) 0.7619
8. (a) 0.2851 (b) 0.1606 (c) 0.7149 (d) 0.2945
9. (a) 0.0627 (b) 0.9373 (c) 0.7638
10. (a) 0.9104 (b) 0.9938 (c) 52.5631
11. (a) 0.9878 (b) 0.8185 (c) 86.16%

    (d) Option 1: Because your score of 81% on this exam represents a Z score of 1.00, which is below the minimum Z score of 1.28, you will not earn an A grade on the exam under this grading option. Option 2: Because your score of 68% on this exam represents a Z score of 2.00, which is well above the minimum Z score of 1.28, you will earn an A grade on the exam under this grading option. You should prefer option 2.

12. (a) 0.4711

    (b) 0.1251

    (c) 8.38 and 16.22

13. The cost of a meal at suburban restaurants is approximately normally distributed because the normal probably plot is approximately a straight line.

14. The cost of attending an NBA basketball game appears to be right-skewed.

15. The alcohol%, calories, and carbohydrates all appear to be right skewed.

References

1. Berenson, M. L., D. M. Levine, and K. A. Szabat. *Basic Business Statistics: Concepts and Applications*, Thirteenth Edition. Upper Saddle River, NJ: Pearson Education, 2015.

2. Levine, D. M., D. Stephan, and K. A. Szabat. *Statistics for Managers Using Microsoft Excel*, Seventh Edition. Upper Saddle River, NJ: Pearson Education, 2014.

3. Levine, D. M., P. P. Ramsey, and R. K. Smidt. *Applied Statistics for Engineers and Scientists Using Microsoft Excel and Minitab*. Upper Saddle River, NJ: Prentice Hall, 2001.

4. Microsoft Excel 2013. Redmond, WA: Microsoft Corporation, 2013.

CHAPTER

6

Sampling Distributions and Confidence Intervals

Inferential statistics relies on the principle that sample statistics can be used to reach conclusions about population parameters. The idea that a small part of a much larger thing, *a sample*, can be used to reach conclusions about that much larger thing, *a population*, at first seems counterintuitive and that causes some people to be dismissive about inferential methods.

This chapter explains the validity of the principle on which inferential statistics relies. You will learn the concept of *statistical* confidence and the statistical methods that enable you to estimate a *confidence interval*.

6.1 Foundational Concepts

Before you can understand the principle that underlies inferential statistics, you must first learn some foundational concepts. These new concepts, when combined with the probability and probability distribution concepts of the previous two chapters, help explain the principle.

All Possible Samples of a Given Sample Size

CONCEPT The set of samples that represents all possible combinations of values from a population for a specific sample size.

EXAMPLES For a population size of 10, the set of all possible samples of $n = 2$ without replacement would be 45 samples.

INTERPRETATION The size of the set of all possible samples of a given sample size can be calculated using combinational mathematics. The set can grow quite large as the population size or sample size increases. For example, if the population size was 100 instead of 10, the set of all possible samples of $n = 2$ without replacement would contain 4,950 different samples.

When examining all possible samples, you work with a sample statistic, such as the mean, for numerical data and the proportion for categorical data, and not the samples themselves. This leads directly to the next concept.

Sampling Distribution

CONCEPT The probability distribution of a sample statistic for all possible samples of a given size n for a given population.

EXAMPLES The sampling distribution of the mean, the sampling distribution of the proportion.

INTERPRETATION Sampling distributions express the variability in the sample statistic computed for each sample, just as a probability distribution expresses that variability for the values of a variable. For example, in the sampling distribution of the mean for a numerical variable, some of the sample means might be smaller than others, some might be larger, and some might be similar to each other. If you knew about this variability in the way that variability is "known" in the probability distributions discussed in Chapter 5, you could use techniques related to those in Chapter 5 to express the likelihood that the sample statistic properly estimates a population parameter. Calculating the sample statistic for all possible samples would be one way but would be impractical for realistic examples. What then?

Central Limit Theorem

CONCEPT The sampling distribution of the mean can be approximated by the normal distribution when the sample size of all possible samples gets large enough.

INTERPRETATION The Central Limit Theorem (CLT) enables you to determine the shape of a sampling distribution without having to resort to the (impractical) method of calculating sample statistics for all samples. What "large enough" means can vary, but, as a general rule, a sample size of 30 will be sufficient for most cases.

The importance of the CLT cannot be overstated. The CLT enables you to apply the knowledge of the normal distribution (see Chapter 5) when analyzing the variability in the sample means. That, in turn, allows you to make statistical judgments such as the ones discussed later in this chapter.

Figure 6.1 on page 116 illustrates that the CLT applies to all types of populations, regardless of their shape. In the figure, the effects of increasing sample size are shown for these populations:

- A normally distributed population (left column)
- A uniformly distributed population in which the values are evenly distributed between the smallest and largest values (middle column)
- An exponentially distributed population in which the values are heavily skewed to the right (right column)

For the normally distributed population, the sampling distribution of the mean is always normally distributed, too. However, as the sample size increases the variability of the sample means decreases, resulting in a narrowing of the graph.

For the other two populations, a **central limiting** effect causes the sample means to become more similar and the shape of the graphs to become more like a normal distribution. This effect happens initially more slowly for the heavily skewed exponential distribution than for the uniform distribution, but when the sample size is increased to 30, the sampling distributions of these two populations converge to the shape of the sampling distribution of the normal population.

Figure 6.1 helps illustrates the following conclusions about the sampling distribution of the mean:

- For most population distributions, regardless of shape, the sampling distribution of the mean is approximately normally distributed if samples of at least 30 observations are selected.
- If the population distribution is fairly symmetrical, the sampling distribution of the mean is approximately normally distributed if samples of at least 15 observations are selected.
- If the population is normally distributed, the sampling distribution of the mean is normally distributed regardless of the sample size.

Figure 6.1

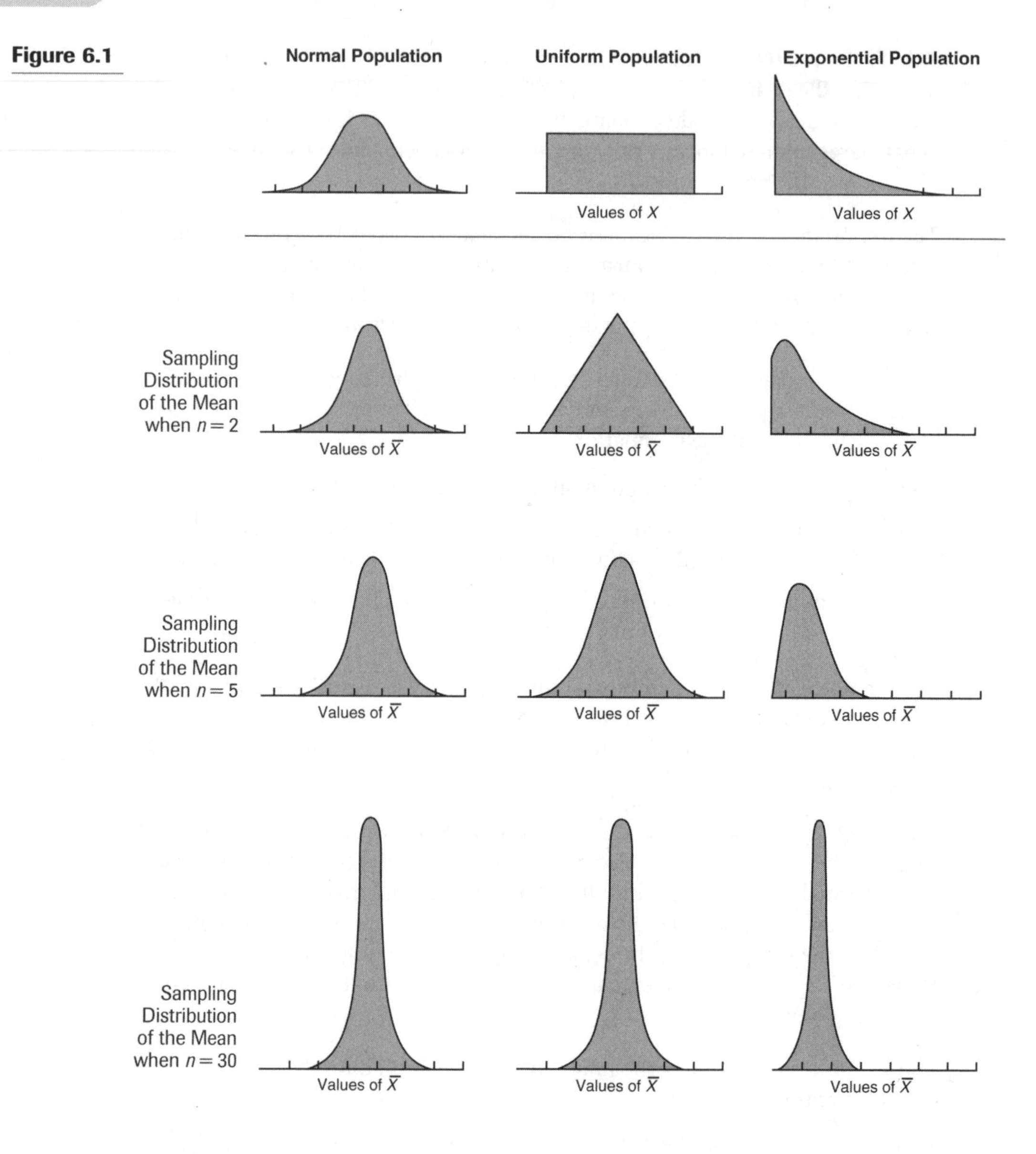

Sampling Distribution of the Proportion

CONCEPT The probability distribution of a proportion for all possible samples of a given size n for a given population.

INTERPRETATION Section 5.2 explains that you can use a binomial distribution to determine probabilities for categorical variables that have only two categories, traditionally labeled "success" and "failure." As the sample size increases for such variables, you can use the normal distribution to approximate the sampling distribution of the number of successes or the proportion of successes.

Specifically, as a general rule, you can use the normal distribution to approximate the binomial distribution when the number of successes and the number of failures are each at least five. For most cases in which you are estimating the proportion, the sample size is more than sufficient to meet the conditions for using the normal approximation.

6.2 Sampling Error and Confidence Intervals

Taking one sample and computing the results of a sample statistic, such as the mean, creates a **point estimate** of the population parameter. This single estimate will almost certainly be different if another sample is selected as WORKED-OUT PROBLEM 1 illustrates.

OrderTimePopulation

WORKED-OUT PROBLEM 1 Take 20 samples of $n = 15$ selected from a population of $N = 200$ order-filling times and compute sample statistics. This population has a population mean = 69.637 and a population standard deviation = 10.411. One such set of 20 samples of $n = 15$ appears in the following table.

SamplesofOrderTimes

Sample	Mean	Standard Deviation	Minimum	Median	Maximum	Range
1	66.12	9.21	47.20	65.00	87.00	39.80
2	73.30	12.48	52.40	71.10	101.10	48.70
3	68.67	10.78	54.00	69.10	85.40	31.40
4	69.95	10.57	54.50	68.00	87.80	33.30
5	73.27	13.56	54.40	71.80	101.10	46.70
6	69.27	10.04	50.10	70.30	85.70	35.60
7	66.75	9.38	52.40	67.30	82.60	30.20
8	68.72	7.62	54.50	68.80	81.50	27.00
9	72.42	9.97	50.10	71.90	88.90	38.80
10	69.25	10.68	51.10	66.50	85.40	34.30
11	72.56	10.60	60.20	69.10	101.10	40.90
12	69.48	11.67	49.10	69.40	97.70	48.60
13	64.65	9.71	47.10	64.10	78.50	31.40
14	68.85	14.42	46.80	69.40	88.10	41.30
15	67.91	8.34	52.40	69.40	79.60	27.20
16	66.22	10.18	51.00	66.40	85.40	34.40
17	68.17	8.18	54.20	66.50	86.10	31.90
18	68.73	8.50	57.70	66.10	84.40	26.70
19	68.57	11.08	47.10	70.40	82.60	35.50
20	75.80	12.49	56.70	77.10	101.10	44.40

From these results, you can observe the following:

- The sample statistics differ from sample to sample. The sample means vary from 64.65 to 75.80, the sample standard deviations vary from 7.62 to 14.42, the sample medians vary from 64.10 to 77.10, and the sample ranges vary from 26.70 to 48.70.
- Some of the sample means are higher than the population mean of 69.637, and some of the sample means are lower than the population mean.
- Some of the sample standard deviations are higher than the population standard deviation of 10.411, and some of the sample standard deviations are lower than the population standard deviation.
- The variation in the sample range from sample to sample is much greater than the variation in the sample standard deviation.

Sample statistics almost always vary from sample to sample. This expected variation is called the **sampling error**.

Sampling Error

CONCEPT The variation that occurs due to selecting a single sample from the population.

EXAMPLE In polls, the plus-or-minus margin of the results; as in "42%, plus or minus 3%, said they were likely to vote for the incumbent."

INTERPRETATION The size of the sampling error is primarily based on the variation in the population itself and on the size of the sample selected. Larger samples have less sampling error, but will be more costly to take.

In practice, only one sample is used as the basis for estimating a population parameter. To account for the differences in the results from sample to sample, you use a **confidence interval estimate**.

Confidence Interval Estimate

CONCEPT A range with an explicit lower and upper limit, stated with a specific degree of certainty, that represents an estimate of a population parameter.

INTERPRETATION As WORKED-OUT PROBLEM 1 illustrates, sample statistics vary and can be less than or greater than the actual population parameter. A confidence interval estimate creates a range centered on a sample statistic and identifies the likelihood that this range includes the actual population parameter. To calculate this range, you need the calculated sample statistic being used to estimate the population parameter and the sampling distribution associated with the sample statistic.

important point

You must state the given degree of certainty, or **confidence**, when reporting an interval estimate, as in "interval estimate with 95% confidence," sometimes more simply phrased as a "95% confidence interval estimate." An interval estimate stated without the degree of certainty is worthless; one stated properly with certainty answers the reservations some have when they wonder how you can reach conclusions about an entire population using only one sample. The answer is that you can, but only with the probabilistic measure that *confidence* represents. An "interval estimate with 95% confidence" means that if all possible samples of the sample size that your single sample uses were taken, 95% percent of interval estimates calculated from those samples would include the population parameter (and 5% would not).

Because you are estimating an interval using one sample and not precisely determining a value, you can never be 100% certain that your interval correctly estimates the population parameter. You trade off the level of confidence and the width of the interval: the greater the confidence that your interval will be correct, the wider the interval will be. In the general case, most settle on 95% confidence as an acceptable trade-off, although 99% confidence (resulting in a wider width) and 90% confidence (resulting in a narrower width) are also used in certain cases.

WORKED-OUT PROBLEM 2 You want to develop 95% confidence interval estimates for the mean from 20 samples of size 15 for the order-filling data presented on page 117. Unlike most real-life problems, the population mean, $\mu = 69.637$, and the population standard deviation, $\sigma = 10.411$, are already known, so the confidence interval estimate for the mean developed from each sample can be compared to the actual value of the population mean.

95% Confidence Interval Estimates from 20 Samples of $n = 15$ Selected from a Population of $N = 200$ with $\mu = 69.637$ and $\sigma = 10.411$

Sample	Mean	Standard Deviation	Lower Limit	Upper Limit
1	66.12	9.21	60.85	71.39
2	73.30	12.48	68.03	78.57
3	68.67	10.78	63.40	73.94
4	69.95	10.57	64.68	75.22
5	73.27	13.56	68.00	78.54
6	69.27	10.04	64.00	74.54
7	66.75	9.38	61.48	72.02
8	68.72	7.62	63.45	73.99
9	72.42	9.97	67.15	77.69
10	69.25	10.68	63.98	74.52
11	72.56	10.60	67.29	77.83
12	69.48	11.67	64.21	74.75
13	64.65	9.71	59.38	69.92
14	68.85	14.42	63.58	74.12
15	67.91	8.34	62.64	73.18
16	66.22	10.18	60.95	71.49
17	68.17	8.18	62.90	73.44
18	68.73	8.50	63.46	74.00
19	68.57	11.08	63.30	73.84
20	75.80	12.49	**70.53**	**81.07**

From the results, you can conclude the following:

- For sample 1, the sample mean is 66.12, the sample standard deviation is 9.21, and the interval estimate for the population mean is 60.85 to 71.39. This enables you to conclude with 95% certainty that the population mean is between 60.85 and 71.39. This is a correct estimate because the population mean of 69.637 is included within this interval.
- Although their sample means and standard deviations differ, the confidence interval estimates for samples 2 through 19 lead to interval estimates that include the population mean value.
- For sample 20, the sample mean is 75.80, the sample standard deviation is 12.49, and the interval estimate for the population mean is 70.53 to 81.07 (highlighted in the results). This is an incorrect estimate because the population mean of 69.637 is not included within this interval.

You might realize that these results are not surprising because the percentage of correct results (19 out of 20) is 95%, equal to the percentage for the confidence interval. However, other specific sets of 20 samples could result in a higher or lower correct percentage that is not exactly 95%. That said, in the long run, 95% of all samples would result in a correct estimate.

6.3 Confidence Interval Estimate for the Mean Using the *t* Distribution (σ Unknown)

Calculating the confidence interval estimate for a mean requires not only knowing the sample mean but the population standard deviation as well. However, in nearly all cases, this parameter is unknown. The ***t* distribution** (see reference 1) allows you to use the sample standard deviation, something that can be always determined, to calculate confidence interval estimates.

t Distribution

CONCEPT The sampling distribution that allows you to develop a confidence interval estimate of the mean using the sample standard deviation.

INTERPRETATION The *t* distribution assumes that the variable being studied comes from a normally distributed population, something that may or may not be known. As a practical matter, though, as long as the sample size is large enough and the population is not very skewed, the *t* distribution can be used to estimate the population mean. However, with a small sample size, or a skewed population distribution, you should verify that the variable does not violate the assumption of **normality**, i.e., that the variable being studied is normally distributed. (You can use a histogram, a box-and-whisker plot, or a normal probability plot to see if the assumption is violated.)

Restaurants

WORKED-OUT PROBLEM 3 Using the cost of a meal per person data for a sample of 50 city restaurants and 50 suburban restaurants collected for WORKED-OUT PROBLEM 7 in Chapter 3 (see page 44), calculate the confidence interval estimate of the mean meal cost.

Spreadsheet results for the confidence interval estimate of the population mean for the mean cost of a meal in city restaurants and in suburban restaurants are as follows:

	A	B	C
1	**Confidence Interval Estimate for the Mean**		
2			
3	**Data**	**City**	**Suburban**
4	**Sample Standard Deviation**	**14.9151**	**11.3785**
5	**Sample Mean**	**49.3**	**44.4**
6	**Sample Size**	**50**	**50**
7	**Confidence Level**	**95%**	**95%**
8			
9	Intermediate Calculations		
10	Standard Error of the Mean	2.1093	1.6092
11	Degrees of Freedom	49	49
12	t Value	2.0096	2.0096
13	Interval Half Width	4.2388	3.2337
14			
15	**Confidence Interval**	**City**	**Suburban**
16	**Interval Lower Limit**	**45.06**	**41.17**
17	**Interval Upper Limit**	**53.54**	**47.63**

To evaluate the assumption of normality necessary to use these estimates, you can use box-and-whisker plots for the city and suburban restaurant meal costs shown on page 55 in Chapter 3. You observe that the box-and-whisker plots have some right-skewness because the tail on the right is longer than the one on the left. However, given the relatively large sample size, you can conclude that any departure from the normality assumption will not seriously affect the validity of the confidence interval estimate.

Based on these results, with 95% confidence, you can conclude that the mean cost of a meal is between \$45.06 and \$53.54 for city restaurants and between \$41.17 and \$47.63 for suburban restaurants.

interested in math?

You use the symbols $\bar{X}$ (sample mean), μ (population mean), S (sample standard deviation), and n (sample size), introduced earlier, and the new symbols α (alpha) and t_{n-1}, which represents the critical value of the t distribution with $n - 1$ degrees of freedom for an area of $\alpha/2$ in the upper tail, to express the confidence interval for the mean as a formula, in cases in which the population standard deviation, σ, is unknown. The symbol alpha, α, is equivalent to 1 minus the confidence percentage. For 95% confidence, α is 0.05 (1–0.95) and $\alpha/2$, the upper tail area, is 0.025.

Using these symbols creates the following equation:

$$\bar{X} \pm t_{n-1}\frac{S}{\sqrt{n}}$$

or expressed as a range:

$$\bar{X} - t_{n-1}\frac{S}{\sqrt{n}} \leq \mu \leq \bar{X} + t_{n-1}\frac{S}{\sqrt{n}}$$

For the city restaurant meal cost example of this section, $\bar{X}$ = 49.3, and S = 14.9151, and because the sample size is 50, there are 49 degrees of freedom. Given 95% confidence, α is 0.05, and the area in the upper tail of the t distribution is 0.025 (0.05/2). Using Table C.2 (in Appendix C), the critical value for the row with 49 degrees of freedom and the column with an area of 0.025 is 2.0096. Substituting these values yields the following result:

$$\bar{X} \pm t_{n-1}\frac{S}{\sqrt{\text{n}}}$$

$$= 49.3 \pm (2.0096)\frac{14.9151}{\sqrt{50}}$$

$$= 49.3 \pm 4.2388$$

$$45.06 \leq \mu \leq 53.54$$

The interval is estimated to be between \$45.06 and \$53.54 with 95% confidence.

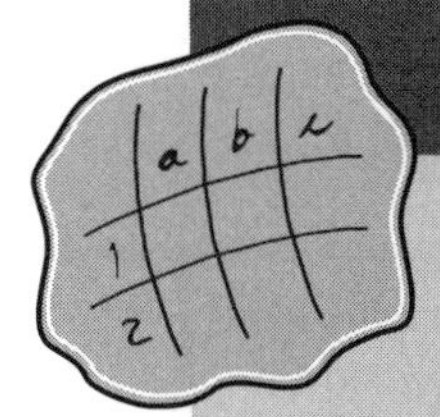

spreadsheet solution

Confidence Interval Estimate for the Mean When σ Is Unknown

Chapter 6 Sigma Unknown contains the spreadsheet similar to the one shown on page 121 that calculates a confidence interval estimate for the population mean when the population standard deviation is unknown. Experiment with this spreadsheet by changing the sample standard deviation, sample mean, sample size, and confidence level in cells B4 through B7.

Best Practices

Use the T.INV.2T(1–*confidence level, degrees of freedom*) function to calculate the critical value of the *t* distribution that is used, in turn, to calculate the confidence interval for the mean when σ is unknown. For a confidence level of 95%, enter 0.05 as the value for **1–*confidence level***.

How-Tos

Tip ADV4 in Appendix E explains how to modify the spreadsheet for use with unsummarized data, such as the data for problems 1 through 3 at the end of this chapter.

6.4 Confidence Interval Estimation for Categorical Variables

For a categorical variable, you can develop a confidence interval to estimate the proportion of successes in a given category.

Confidence Interval Estimation for the Proportion

CONCEPT The sampling distribution of the proportion that allows you to develop a confidence interval estimate of the proportion using the sample proportion of successes, *p*. The sample statistic *p* follows a binomial distribution that can be approximated by the normal distribution for most studies.

EXAMPLE The proportion of voters who would vote for a certain candidate in an election, the proportion of consumers who own a particular brand of smartphone, the proportion of medical tests in a hospital that need to be repeated.

INTERPRETATION This type of confidence interval estimate uses the sample proportion of successes, p, equal to the number of successes divided by the sample size, to estimate the population proportion. (Categorical variables have no population means.)

For a given sample size, confidence intervals for proportions are wider than those for numerical variables. With continuous variables, the measurement on each respondent contributes more information than for a categorical variable. In other words, a categorical variable with only two possible values is a very crude measure compared with a continuous variable, so each observation contributes only a little information about the parameter being estimated.

WORKED-OUT PROBLEM 4 According to a Pew Research Center report, 463 out of 1,006 Internet-using adults polled said that the Internet would be very hard or impossible to give up. (Source: "The Web at 25 in the U.S.," **bit.ly/1dE8jFV**.) Using this sample, you want to estimate the proportion of all Internet-using adults who say that the Internet would be hard or impossible to give up.

Spreadsheet results for the confidence interval estimate of the population proportion, using 95% confidence, are as follows:

	A	B
1	**Confidence Interval Estimate for the Proportion**	
2		
3	**Data**	
4	**Sample Size**	**1006**
5	**Number of Successes**	**463**
6	**Confidence Level**	**95%**
7		
8	Intermediate Calculations	
9	Sample Proportion	0.4602
10	Z Value	-1.9600
11	Standard Error of the Proportion	0.0157
12	Interval Half Width	0.0308
13		
14	**Confidence Interval**	
15	**Interval Lower Limit**	**0.4294**
16	**Interval Upper Limit**	**0.4910**

Based on these results, you estimate that between 42.94% and 49.10% of all Internet-using adults would say that the Internet would be hard or impossible to give up.

interested in math?

You use the symbols p (sample proportion of success), n (sample size), and Z (Z score), previously introduced, and the symbol π for the population proportion, to assemble the equation for the confidence interval estimate for the proportion:

$$p \pm Z\sqrt{\frac{p(1-p)}{n}}$$

or expressed as a range:

$$p - Z\sqrt{\frac{p(1-p)}{n}} \leq \pi \leq p + Z\sqrt{\frac{p(1-p)}{n}}$$

Z = Critical value from the normal distribution

For WORKED-OUT PROBLEM 4, n = 1,006 and p = 463/1,006 = 0.4602. For a 95% level of confidence, the lower tail area of 0.025 provides a Z value from the normal distribution of –1.96, and the upper tail area of 0.025 provides a Z value from the normal distribution of +1.96. Substituting these numbers into the preceding equation yields the following result:

$$p \pm Z\sqrt{\frac{p(1-p)}{n}}$$

$$= 0.4602 \pm (1.96)\sqrt{\frac{(0.4602)(0.5398)}{1{,}006}}$$

$$= 0.4602 \pm (1.96)((0.0157)$$

$$= 0.4602 \pm 0.0308$$

$$0.4294 \leq \pi \leq 0.4910$$

The proportion of Internet-using adults who say that the Internet would be hard or impossible to give up is estimated to be between 42.94% and 49.10%.

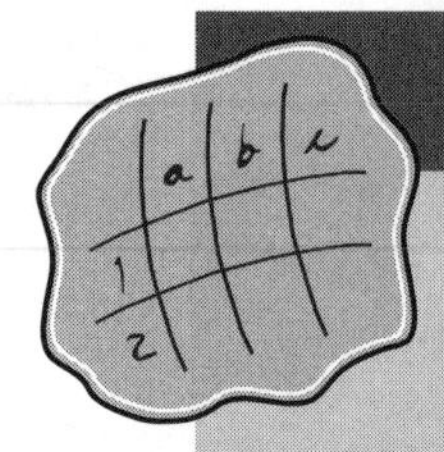

spreadsheet solution

Confidence Interval Estimate for the Proportion

Chapter 6 Proportion contains the spreadsheet shown on page 124 that calculates a confidence interval estimate for the population proportion. Experiment with this spreadsheet by changing the sample size, number of successes, and confidence level in cells B4 through B6.

Best Practices

Use the NORM.S.INV((1–*confidence level*)/2) function to calculate the critical value of the normal distribution that is used, in turn, to calculate the confidence interval for the proportion. For a confidence level of 95%, enter 0.05 as the value for 1–*confidence level*.

How-Tos

Use the absolute function ABS, used in cell B12 of **Chapter 6 Proportion**, to help calculate the absolute value of the half-width of the confidence interval.

6.5 Bootstrapping Estimation

When you cannot assume that a population is normally distributed, you cannot use the confidence interval estimation methods for population parameters discussed in earlier sections of this chapter. You can, however, use **bootstrapping estimation** to estimate a population parameter.

Bootstrapping

CONCEPT The confidence interval estimation methods that use repeated *resampling with replacement* of the initial sample being used as the basis of estimation.

INTERPRETATION Because bootstrap methods are based on the initial sample, not on the population, bootstrap methods do not need to make assumptions, such as normality, about the nature of the population distribution. Bootstrapping consists of a variety of methods, all of which rely on repeated resampling. An example of a bootstrap method to estimate the population mean consists of the following steps:

1. Select a random sample of size *n* *without replacement* from a population of size *N*.
2. Resample the initial sample by selecting *n* values *with replacement* from the *n* values in the initial sample, and compute the sample means for this resample.
3. Repeat step 2 *m* number of times to produce *m* resamples.
4. Construct the resampling distribution of the sample mean from each of the *m* samples.
5. Construct an ordered array of the entire set of resampled means.
6. Find the values that exclude the smallest $\alpha/2 \times 100\%$ of means and the largest $\alpha/2 \times 100\%$ of means. These values become the lower and upper limits of the bootstrap confidence interval estimate of the population mean with $(1-\alpha)\%$ confidence.

When applying a bootstrapping estimation method, you typically use specialized statistical software to select a very large number of resamples and to perform the bootstrapping estimation. What "very large" means keeps increasing as the power of computing technology increases (the more resamples, the less sampling error but the greater the processing time).

WORKED-OUT PROBLEM 5 An insurance company seeks to reduce the amount of time it takes to approve life insurance applications. You collect data by selecting a random sample of 27 approved policies during a period of one month. This sample contains the following values for total processing times (in days):

Insurance

73	19	16	64	28	28	31	90	60	56	31	56	22	18
45	48	17	17	17	91	92	63	50	51	69	16	17	

A 90% confidence interval estimate for the population mean processing time indicates that the mean processing time for the population of life insurance applications is between 35.59 and 52.19 days. This estimate assumes that the population of processing times is normally distributed, but a box-and-whisker plot and a normal probability plot (not shown) indicate that the population is right-skewed. This raises questions about the validity of the confidence interval, and you decide to use a bootstrap estimation method that will use 100 resamples.

The following list shows the first resample that is based on the initial sample of 27 processing times shown earlier:

16	16	16	17	17	17	17	17	19	22	28	31	31	48
51	56	56	60	60	64	64	64	69	**73**	**73**	90	92	

You note that this first resample omits some values (18, 45, 50, 63, and 91) that appear in the initial sample and repeats some other values more times than they appear in the initial sample. For example, the value 73 (highlighted) appears only once in the initial sample but twice in the resample as a result of resampling *with replacement*. After taking 100 resamples of the processing times, you compute sample means and construct the following ordered array of the sample means for the 100 resamples:

31.5926	33.9259	35.4074	36.5185	**36.6296**	36.9630	37.0370	37.0741
37.1481	37.3704	37.9259	38.1111	38.1481	38.2222	38.2963	38.7407
38.8148	38.8519	38.8889	39.0000	39.1852	39.3333	39.3704	39.6667
40.1481	40.5185	40.6296	40.9259	40.9630	41.2593	41.2963	41.7037
41.8889	42.0741	42.1111	42.1852	42.8519	43.0741	43.1852	43.3704
43.4444	43.7037	43.8148	43.8519	43.8519	43.9259	43.9630	44.1481
44.4074	44.5556	44.7778	45.0000	45.4444	45.5185	45.5556	45.6667
45.7407	45.8519	45.9630	45.9630	46.0000	46.1111	46.2963	46.2963
46.3333	46.3333	46.4815	46.6667	46.7407	46.9630	47.0741	47.2222
47.2963	47.3704	47.4815	47.4815	47.5556	47.6667	47.8519	48.5185
48.8889	49.0000	49.2222	49.4444	49.4815	49.4815	49.6296	49.6296
49.7407	50.2963	50.4074	50.5926	50.9259	51.4074	51.4815	**51.5926**
51.9259	52.3704	53.4074	54.3333				

You need to identify the fifth-smallest and fifth-largest values to exclude the smallest and largest 5% of the resample means (90% confidence is equal to $\alpha = 0.1$; $\alpha/2 = 0.05$, and $0.05 \times 100\% = 5$). In the ordered array, the fifth-smallest value is 36.6296 and the fifth-largest value is 51.5926 (highlighted). Therefore, the 90% bootstrap confidence interval estimate of the population mean processing time is 36.6296 to 51.5926 days.

Important Equations

Confidence interval for the mean with σ unknown:

$$\bar{X} \pm t_{n-1} \frac{S}{\sqrt{n}}$$

(6.1) *or*

$$\bar{X} - t_{n-1} \frac{S}{\sqrt{n}} \leq \mu \leq \bar{X} + t_{n-1} \frac{S}{\sqrt{n}}$$

Confidence interval estimate for the proportion:

$$p \pm Z\sqrt{\frac{p(1-p)}{n}}$$

(6.2) *or*

$$p - Z\sqrt{\frac{p(1-p)}{n}} \le \pi \le p + Z\sqrt{\frac{p(1-p)}{n}}$$

One-Minute Summary

Which confidence interval estimate you use depends on the type of variable being studied:

- Use the confidence interval estimate for the mean for a numerical variable.
- Use the confidence interval estimate for the proportion for a categorical variable.

Test Yourself

Short Answers

1. The sampling distribution of the mean can be approximated by the normal distribution:
   (a) as the number of samples gets "large enough"
   (b) as the sample size (number of observations in each sample) gets large enough
   (c) as the size of the population standard deviation increases
   (d) as the size of the sample standard deviation decreases
2. The sampling distribution of the mean requires ________ sample size to reach a normal distribution if the population is skewed than if the population is symmetrical.
   (a) the same
   (b) a smaller
   (c) a larger
   (d) The two distributions cannot be compared.
3. Which of the following is true regarding the sampling distribution of the mean for a large sample size?
   (a) It has the same shape and mean as the population.
   (b) It has a normal distribution with the same mean as the population.
   (c) It has a normal distribution with a different mean from the population.

4. For samples of $n = 30$, for most populations, the sampling distribution of the mean will be approximately normally distributed:
   (a) regardless of the shape of the population
   (b) if the shape of the population is symmetrical
   (c) if the standard deviation of the mean is known
   (d) if the population is normally distributed

5. For samples of $n = 1$, the sampling distribution of the mean will be normally distributed:
   (a) regardless of the shape of the population
   (b) if the shape of the population is symmetrical
   (c) if the standard deviation of the mean is known
   (d) if the population is normally distributed

6. A 99% confidence interval estimate can be interpreted to mean that:
   (a) If all possible samples are taken and confidence interval estimates are developed, 99% of them would include the true population mean somewhere within their interval.
   (b) You have 99% confidence that you have selected a sample whose interval does include the population mean.
   (c) Both a and b are true.
   (d) Neither a nor b is true.

7. Which of the following statements is false?
   (a) There is a different critical value for each level of alpha (α).
   (b) Alpha (α) is the proportion in the tails of the distribution that is outside the confidence interval.
   (c) You can construct a 100% confidence interval estimate of μ.
   (d) In practice, the population mean is the unknown quantity that is to be estimated.

8. Sampling distributions describe the distribution of:
   (a) parameters
   (b) statistics
   (c) both parameters and statistics
   (d) neither parameters nor statistics

9. In the construction of confidence intervals, if all other quantities are unchanged, an increase in the sample size will lead to a ________ interval.
   (a) narrower
   (b) wider
   (c) less significant
   (d) the same

10. As an aid to the establishment of personnel requirements, the manager of a bank wants to estimate the mean number of people who arrive at the bank during the two-hour lunch period from 12 noon to 2 p.m. The director randomly selects 64 different two-hour lunch periods from 12 noon to 2 p. m. and determines the number of people who arrive for each. For this sample, $\bar{X} = 49.8$ and $S = 5$. Which of the following assumptions is necessary in order for a confidence interval to be valid?
    (a) The population sampled from has an approximate normal distribution.
    (b) The population sampled from has an approximate t distribution.
    (c) The mean of the sample equals the mean of the population.
    (d) None of these assumptions are necessary.
11. A university dean is interested in determining the proportion of students who are planning to attend graduate school. Rather than examine the records for all students, the dean randomly selects 200 students and finds that 118 of them are planning to attend graduate school. The 95% confidence interval for p is 0.59 ± 0.07. Interpret this interval.
    (a) You are 95% confident that the true proportion of all students planning to attend graduate school is between 0.52 and 0.66.
    (b) There is a 95% chance of selecting a sample that finds that between 52% and 66% of the students are planning to attend graduate school.
    (c) You are 95% confident that between 52% and 66% of the sampled students are planning to attend graduate school.
    (d) You are 95% confident that 59% of the students are planning to attend graduate school.
12. In estimating the population mean with the population standard deviation unknown, if the sample size is 12, there will be _____ degrees of freedom.
13. The Central Limit Theorem is important in statistics because
    (a) It states that the population will always be approximately normally distributed.
    (b) It states that the sampling distribution of the sample mean is approximately normally distributed for a large sample size n regardless of the shape of the population.
    (c) It states that the sampling distribution of the sample mean is approximately normally distributed for any population regardless of the sample size.
    (d) For any sized sample, it says the sampling distribution of the sample mean is approximately normal.

14. For samples of $n = 15$, the sampling distribution of the mean will be normally distributed:
    (a) regardless of the shape of the population
    (b) if the shape of the population is symmetrical
    (c) if the standard deviation of the mean is known
    (d) if the population is normally distributed

Answer True or False:

15. Other things being equal, as the confidence level for a confidence interval increases, the width of the interval increases.
16. As the sample size increases, the effect of an extreme value on the sample mean becomes smaller.
17. A sampling distribution is defined as the probability distribution of possible sample sizes that can be observed from a given population.
18. The t distribution is used to construct confidence intervals for the population mean when the population standard deviation is unknown.
19. In the construction of confidence intervals, if all other quantities are unchanged, an increase in the sample size will lead to a wider interval.
20. The confidence interval estimate that is constructed will always correctly estimate the population parameter.

Answers to Test Yourself Short Answers

1. b
2. c
3. b
4. a
5. d
6. c
7. c
8. b
9. a
10. d
11. a
12. 11
13. b
14. b
15. True
16. True
17. False
18. True
19. False
20. False

Problems

OnlinePrices

1. How do the prices at online grocers compare to the prices at supermarkets? A basket of products consisting of Kellogg's Raisin Bran, a half-gallon of 2% milk, Starbucks coffee, boneless chicken breasts, Sprite soda, Lay's potato chips, 5 bananas, Barilla spaghetti, a pint of Ben & Jerry's ice cream, Old Spice body wash, 2 lemons, Fig Newtons, Glad

ClingWrap, and seedless grapes was purchased at online grocers, with the following results:

Company	Total Cost
Peapod (New York)	72.95
FreshDirect	75.13
Safeway.com	75.85
AmazonFresh	62.13
Walmart.com	52.70
Instacart	72.19
Peapod (Indianapolis)	70.57

Source: Data extracted from G. A. Fowler, "Price Check: Do Online Grocers Beat Supermarkets?" *The Wall Street Journal*, 8 January 2014, p. D1, D2.

Construct a 95% confidence interval estimate of the mean price of the basket of products at online grocers.

Protein

2. The following table contains calories, protein, percentage of calories from fat, percentage of calories from saturated fat, and cholesterol of popular protein foods (fresh red meats, poultry, and fish).

Food	Calories	Protein	Pctage of Calories from Fat	Pctage of Calories from Saturated Fat	Cholesterol
Beef, ground, extra lean	250	25	58	23	82
Beef, ground, regular	287	23	66	26	87
Beef, round	184	28	24	12	82
Brisket	263	28	54	21	91
Flank steak	244	28	51	22	71
Lamb leg roast	191	28	38	16	89
Lamb loin chop, broiled	215	30	42	17	94
Liver, fried	217	27	36	12	482
Pork loin roast	240	27	52	18	90
Sirloin	208	30	37	15	89
Spareribs	397	29	67	27	121
Veal cutlet, fried	183	33	42	20	127
Veal rib roast	175	26	37	15	131
Chicken, with skin, roasted	239	27	51	14	88
Chicken, no skin, roast	190	29	37	10	89
Turkey, light meat, no skin	157	30	18	6	69

Food	Calories	Protein	Pctage of Calories from Fat	Pctage of Calories from Saturated Fat	Cholesterol
Clams	98	16	6	0	39
Cod	98	22	8	1	74
Flounder	99	21	12	2	54
Mackerel	199	27	77	20	100
Ocean perch	110	23	13	3	53
Salmon	182	27	24	5	93
Scallops	112	23	8	1	56
Shrimp	116	24	15	2	156
Tuna	181	32	41	10	48

Source: U.S. Department of Agriculture.

Construct 95% confidence interval estimates of the mean calories, protein, percentage of calories from fat, percentage of calories from saturated fat, and cholesterol of the popular protein foods.

DomesticBeer

3. The **DomesticBeer** file contains the percentage alcohol, number of calories per 12 ounces, and number of carbohydrates (in grams) per 12 ounces for 152 of the best-selling domestic beers in the United States. (Data extracted from **www.beer100.com/beercalories.htm**, 20 March 2013.)
   a. Construct 95% confidence interval estimates of the mean percentage alcohol, mean number of calories per 12 ounces, and mean number of carbohydrates (in grams) per 12 ounces.
   b. Do you need to assume that the variables in (a) are normally distributed in order to construct the confidence intervals in (a)?
4. In a survey of 2,046 shoppers, 1,391 said that one bad experience would cause them to shun a retailer. (Source: "Snapshots," *USA Today*, 29 November 2013, p. 1D.) Construct a 95% confidence interval estimate of the proportion of all shoppers who would shun a retailer based on one bad shopping experience.
5. In a survey of 792 Internet and smartphone users, 681 said that they had taken steps online to remove or mask their digital footprints. (Source: E. Dwoskin, "Give Me Back My Privacy, *The Wall Street Journal*, 24 March 2014, p. R2.) Construct a 95% confidence interval estimate of the proportion of all Internet and smartphone users who would take steps online to remove or mask their digital footprints.

6. In a survey of 790 adult smartphone users, 198 said that they secretly tried to use their smartphone in a public restroom. (Source: "Snapshots," *USA Today*, 16 January 2014, p. 1D.) Construct a 95% confidence interval estimate of the proportion of all adult smartphone users who secretly tried to use their smartphone in a public restroom.

Answers to Test Yourself Problems

1. $61.00 ≤ μ ≤ $76.58
2. Calories: 164.80 ≤ μ ≤ 222.00; protein: 24.97 ≤ μ ≤ 28.07%; calories from fat: 28.23 ≤ μ ≤ 44.89%; calories from saturated fat: 9.25 ≤ μ ≤ 16.19; cholesterol: 67.68 ≤ μ ≤ 136.72.
3. (a) Alcohol%: 5.01 ≤ μ ≤ 5.46; calories: 147.16 ≤ μ ≤ 161.45; carbohydrates: 11.18 ≤ μ ≤ 12.75.

   (b) The sample size is large ($n = 152$), so the use of the t distribution to construct the confidence interval is appropriate because the validity will not be affected.
4. 0.6596 ≤ π ≤ 0.7001
5. 0.8357 ≤ π ≤ 0.8840
6. 0.2204 ≤ π ≤ 0.2809

References

1. Berenson, M. L., D. M. Levine, and K. A. Szabat. *Basic Business Statistics: Concepts and Applications, Thirteenth Edition*. Upper Saddle River, NJ: Pearson Education, 2015.
2. Cochran, W. G. *Sampling Techniques, Third Edition*. New York: Wiley, 1977.
3. Diaconis, P. and B. Efron. "Computer-Intensive Methods in Statistics." *Scientific American*, 248, 1983, pp. 116–130.
4. Efron, B., and R. Tibshirani. *An Introduction to the Bootstrap*. Boca Raton, FL: Chapman and Hall/CRC, 1995.
5. Gunter, B. "Bootstrapping: How to Make Something from Almost Nothing and Get Statistically Valid Answers Part I: Brave New World." *Quality Progress*, 24, December 1991, pp. 97–103.

6. Levine, D. M., D. Stephan, and K. A. Szabat. *Statistics for Managers Using Microsoft Excel, Seventh Edition*. Upper Saddle River, NJ: Pearson Education, 2014.

7. Levine, D. M., P. P. Ramsey, and R. K. Smidt, *Applied Statistics for Engineers and Scientists Using Microsoft Excel and Minitab*. Upper Saddle River, NJ: Prentice Hall, 2001.

8. Microsoft Excel 2013. Redmond, WA: Microsoft Corporation, 2012.

9. Varian, H. “Bootstrap Tutorial.” *Mathematica Journal*, 2005, 9, pp. 768–775.

CHAPTER 7

Fundamentals of Hypothesis Testing

Science progresses by first stating tentative explanations, or hypotheses, about natural phenomena and then by proving (or disproving) those hypotheses through investigation and testing. Statisticians have adapted this scientific method by developing an inferential method called **hypothesis testing** that evaluates a claim made about the value of a population parameter by using a sample statistic. In this chapter, you learn the basic concepts and principles of hypothesis testing and the statistical assumptions necessary for performing hypothesis testing.

7.1 The Null and Alternative Hypotheses

Unlike the broader hypothesis testing of science, statistical hypothesis testing always involves evaluating a claim made about the value of a population parameter. This claim is stated as a pair of statements: the null hypothesis and the alternative hypothesis.

Null Hypothesis

CONCEPT The statement that a population parameter is equal to a specific value or that the population parameters from two or more groups are equal.

EXAMPLES "The population mean time to answer customer emails was 4 hours last year," "the mean height for women is the same as the mean height for men," "at a restaurant, the proportion of orders filled correctly for drive-through customers is the same as the proportion of orders filled correctly for sit-down customers."

important point

INTERPRETATION The null hypothesis always expresses an equality and is always paired with another statement, the alternative hypothesis. A null hypothesis is considered true until evidence indicates otherwise. If you can conclude that the null hypothesis is false, then the alternative hypothesis must be true.

You use the symbol H_0 to identify the null hypothesis and write a null hypothesis using an equal sign and the symbol for the population parameter, as in H_0: $\mu = 4$ or H_0: $\mu_1 = \mu_2$ or H_0: $\pi_1 = \pi_2$. (Remember that in statistics, the symbol π represents the population proportion and not the ratio of the circumference to the diameter of a circle, as the symbol represents in geometry.)

Alternative Hypothesis

CONCEPT The statement paired with a null hypothesis that is mutually exclusive to the null hypothesis.

EXAMPLES "The population mean for the time to answer customer emails was not 4 hours last year" (which would be paired with the example for the null hypothesis in the preceding section); "the mean height for women is not the same as the mean height for men" (paired with the second example for the null hypothesis); "at a restaurant, the proportion of food orders filled correctly for drive-through customers is not the same as the proportion of food orders filled correctly for sit-down customers" (paired with the third example for the null hypothesis).

INTERPRETATION The alternative hypothesis is typically the idea you are studying concerning your data. The alternative hypothesis always expresses an inequality, either between a population parameter and a specific value or between two or more population parameters and is always paired with the null hypothesis. You use the symbol H_1 to identify the alternative hypothesis and write an alternative hypothesis using either a not-equal sign or a less than or greater than sign, along with the symbol for the population parameter, as in H_1: $\mu \neq 4$ or H_1: $\mu_1 \neq \mu_2$ or H_0: $\pi_1 \neq \pi_2$.

The alternative hypothesis represents the conclusion reached by rejecting the null hypothesis. You reject the null hypothesis if evidence from the sample statistic indicates that the null hypothesis is unlikely to be true. However, if you cannot reject the null hypothesis, you cannot claim to have proven the null hypothesis. Failure to reject the null hypothesis means (only) that you have failed to prove the alternative hypothesis.

7.2 Hypothesis Testing Issues

In hypothesis testing, you use the sample statistic to estimate the population parameter stated in the null hypothesis. For example, to evaluate the null hypothesis "the population mean time to answer customer emails was 4 hours last year," you would use the sample mean time to estimate the population mean time. As Chapter 6 establishes, a sample statistic is unlikely to be identical to its corresponding population parameter, and in that chapter, you learned to construct an interval estimate for the parameter based on the statistic.

If the sample statistic is not the same as the population parameter, as it almost never is, the issue of whether to reject the null hypothesis involves deciding how different the sample statistic is from its corresponding population parameter. (In the case of two groups, the issue can be expressed, under certain conditions, as deciding how different the sample statistics of each group are to each other.)

Without a rigorous procedure that includes a clear definition of a difference, you would find it hard to decide on a consistent basis whether a null hypothesis is false and, therefore, whether to reject or not reject the null hypothesis. Statistical hypothesis-testing methods provide such definitions and enable you to restate the decision-making process as the probability of computing a given sample statistic, if the null hypothesis were true through the use of a test statistic and a risk factor.

Test Statistic

CONCEPT The value based on the sample statistic and the sampling distribution for the sample statistic.

EXAMPLES Test statistic for the difference between two sample means (Chapter 8), test statistic for the difference between two sample proportions (Chapter 8), test statistic for the difference between the means of more than two groups (Chapter 9), test statistic for the slope (Chapter 10).

INTERPRETATION If you are testing whether the mean of a population was equal to a specific value, the sample statistic is the sample mean. The test statistic is based on the difference between the sample mean and the value of the population mean stated in the null hypothesis. This test statistic follows a statistical distribution called the *t* distribution that is discussed in Sections 8.2 and 8.3.

If you are testing whether the mean of population one is equal to the mean of population two, the sample statistic is the difference between the mean in sample one and the mean in sample two. The test statistic is based on the difference between the mean in sample one and the mean in sample two. Under certain circumstances, this test statistic also follows the *t* distribution.

The sampling distribution of the test statistic is divided into two regions, a **region of rejection** (also known as the critical region) and a **region of nonrejection**. If the test statistic falls into the region of nonrejection, the null hypothesis is not rejected.

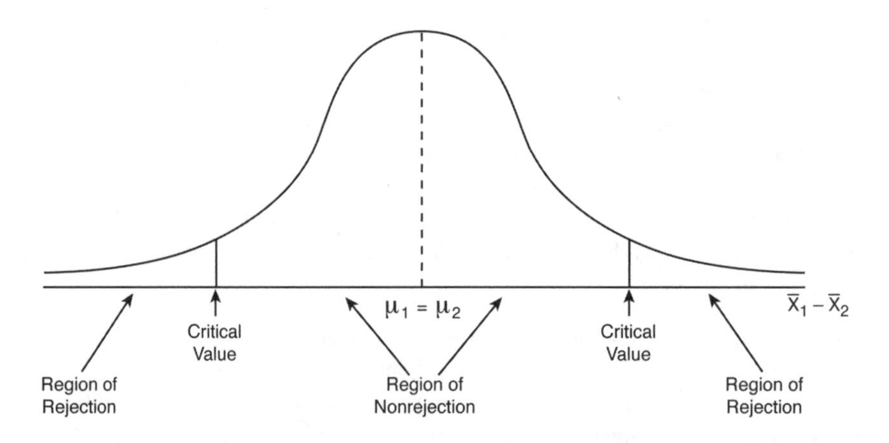

The region of rejection contains the values of the test statistic that are unlikely to occur if the null hypothesis is true. If the null hypothesis is false, these values are likely to occur. Therefore, if you observe a value of the test statistic that falls into the rejection region, you reject the null hypothesis, because that value is unlikely if the null hypothesis is true.

To make a decision concerning the null hypothesis, you first determine the critical value of the test statistic that separates the nonrejection region from the rejection region. You determine the critical value by using the appropriate sampling distribution and deciding on the risk you are willing to take of rejecting the null hypothesis when it is true.

Practical Significance Versus Statistical Significance

important point

Another issue in hypothesis testing concerns the distinction between a statistically significant difference and a practical significant difference. Given a large enough sample size, it is always possible to detect a statistically significant difference. This is because no two things in nature are exactly equal. So, with a large enough sample size, you can always detect the natural difference between two populations. You need to be aware of the real-world practical implications of the statistical significance.

7.3 Decision-Making Risks

In hypothesis testing, you always face the possibility that either you will wrongly reject the null hypothesis or wrongly not reject the null hypothesis. These possibilities are called type I and type II errors, respectively.

Type I Error

CONCEPT The error that occurs if the null hypothesis H_0 is rejected when it is true and should not be rejected.

INTERPRETATION The risk, or probability, of a type I error occurring is identified by the Greek lowercase alpha, α. Alpha is also known as the level of significance of the statistical test. Traditionally, you control the probability of a type I error by deciding the risk level α you are willing to tolerate of rejecting the null hypothesis when it is true. Because you specify the level of significance before performing the hypothesis test, the risk of committing a type I error, α, is directly under your control. The most common α values are 0.01, 0.05, and 0.10, and researchers traditionally select a value of 0.05 or smaller.

When you specify the value for α, you determine the rejection region, and using the appropriate sampling distribution, the critical value or values that divide the rejection and nonrejection regions are determined.

Type II Error

CONCEPT The error that occurs if the null hypothesis H_0 is not rejected when it is false and should be rejected.

INTERPRETATION The risk, or probability, of a type II error occurring is identified by the Greek lowercase beta, β. The probability of a type II error depends on the size of the difference between the value of the population parameter stated in the null hypothesis and the actual population value. Unlike the type I error, the type II error is not directly established by you. Because large differences are easier to find, as the difference between the value of the population parameter stated in the null hypothesis and its corresponding population parameter increases, the probability of a type II error decreases. Therefore, if the difference between the value of the population parameter stated in the null hypothesis and the corresponding parameter is small, the probability of a type II error will be large.

The arithmetic complement of beta, $1 - \beta$, is known as the **power of the test** and represents the probability of rejecting the null hypothesis when it is false and should be rejected.

Risk Trade-Off

The types of errors and their associated risks are summarized in Table 7.1. The probabilities of the two types of errors have an inverse relationship. When you decrease α, you always increase β, and when you decrease β, you always increase α.

TABLE 7.1

Risks and Decisions in Hypothesis Testing

		Actual Situation	
		H_0 True	**H_0 False**
Statistical Decision	**Do not reject H_0**	Correct decision Confidence $= 1 - \alpha$	Type II error P(Type II error) $= \beta$
	Reject H_0	Type I error P(Type I error) $= \alpha$	Correct decision Power $= 1 - \beta$

One way in which you can lower β without affecting the value of α is to increase sample size. Larger sample sizes generally permit you to detect even very small differences between the hypothesized and actual values of the population parameter. For a given level of α, increasing the sample size will decrease β and therefore increase the power of the test to detect that the null hypothesis H_0 is false.

In establishing a value for α, you need to consider the negative consequences of a type I error. If these consequences are substantial, you can set $\alpha = 0.01$ instead of 0.05 and tolerate the greater β that results. If the negative consequences of a type II error most concern you, you can select a larger value for α (for example, 0.05 rather than 0.01) and benefit from the lower β that you will have.

7.4 Performing Hypothesis Testing

When you perform a hypothesis test, you should follow the steps of hypothesis testing in this order:

1. State the null hypothesis, H_0, and the alternative hypothesis, H_1.
2. Evaluate the risks of making type I and II errors, and choose the level of significance, α, and the sample size as appropriate.
3. Determine the appropriate test statistic and sampling distribution to use and identify the critical values that divide the rejection and nonrejection regions.
4. Collect the data, calculate the appropriate test statistic, and determine whether the test statistic has fallen into the rejection or the nonrejection region.
5. Make the proper statistical inference. Reject the null hypothesis if the test statistic falls into the rejection region. Do not reject the null hypothesis if the test statistic falls into the nonrejection region.

The *p*-Value Approach to Hypothesis Testing

Most modern statistical software, including the functions found in spreadsheet programs and calculators, can calculate the probability value known as the *p*-value that you can also use to determine whether to reject the null hypothesis.

p-Value

CONCEPT The probability of computing a test statistic equal to or more extreme than the sample results, given that the null hypothesis H_0 is true.

INTERPRETATION The *p*-value is the smallest level at which H_0 can be rejected for a given set of data. You can consider the *p*-value the actual risk of having a type I error for a given set of data. Using *p*-values, the decision rules for rejecting the null hypothesis are as follows:

important point

- If the *p*-value is greater than or equal to α, do not reject the null hypothesis.
- If the *p*-value is less than α, reject the null hypothesis.
- Many people confuse this rule, mistakenly believing that a high *p*-value is reason for rejection. You can avoid this confusion by remembering the following saying:

  "If the *p*-value is low, then H_0 must go."

In practice, most researchers today use *p*-values for several reasons, including efficiency of the presentation of results. The *p*-value is also known as the **observed level of significance**. When using *p*-values, you can restate the steps of hypothesis testing as follows:

1. State the null hypothesis, H_0, and the alternative hypothesis, H_1.
2. Evaluate the risks of making type I and II errors, and choose the level of significance, α, and the sample size as appropriate.
3. Collect the data and calculate the sample value of the appropriate test statistic.
4. Calculate the *p*-value based on the test statistic and compare the *p*-value to α.
5. Make the proper statistical inference. Reject the null hypothesis if the *p*-value is less than α. Do not reject the null hypothesis if the *p*-value is greater than or equal to α.

7.5 Types of Hypothesis Tests

Your choice of which statistical test to use when performing hypothesis testing is influenced by the following factors:

- Number of groups of data: one, two, or more than two
- Relationship stated in alternative hypothesis H_1: not equal to or inequality (less than, greater than)
- Type of variable (population parameter): numerical (mean) or categorical (proportion)

Number of Groups

One group of hypothesis tests, more formally known as one-sample tests, are of limited practical use, because if you are interested in examining the value of a population parameter, you can usually use one of the confidence interval estimate methods of Chapter 6. Two-sample tests, examining the differences between two groups, can be found in the WORKED-OUT PROBLEMS of Sections 8.1 through 8.3. Tests for more than two groups are discussed in Chapter 9.

Relationship Stated in Alternative Hypothesis H_1

Alternative hypotheses can be stated either using the not-equal sign, as in, H_1: $\mu_1 \neq \mu_2$; or by using an inequality, such as H_1: $\mu_1 > \mu_2$. You use a **two-tail test** for alternative hypotheses that use the not-equal sign and use a **one-tail test** for alternative hypotheses that contain an inequality.

One-tail and two-tail test procedures are very similar and differ mainly in the way they use critical values to determine the region of rejection. Throughout this book, two-tail hypothesis tests are featured. One-tail tests are not further discussed in this book, although WORKED-OUT PROBLEM 8 of Chapter 8 on page 167 illustrates one possible use for such tests.

Type of Variable

The type of variable, numerical or categorical, also influences the choice of hypothesis test used. For a numerical variable, the test might examine the population mean or the differences among the means, if two or more groups are used. For a categorical variable, the test might examine the population proportion or the differences among the population proportions if two or more groups are used. Tests involving two groups for each type of variable can be found in the WORKED-OUT PROBLEMS of Sections 8.1 through 8.3. Tests involving more than two groups for each type of variable are featured in Chapter 9.

One-Minute Summary

Hypotheses

- Null hypothesis
- Alternative hypothesis

Types of errors

- Type I error
- Type II error

Hypothesis testing approach

- Test statistic
- *p*-value

Hypothesis test relationship

- One-tail test
- Two-tail test

Test Yourself

1. A type II error is committed when:
   (a) you reject a null hypothesis that is true
   (b) you don't reject a null hypothesis that is true
   (c) you reject a null hypothesis that is false
   (d) you don't reject a null hypothesis that is false
2. A type I error is committed when:
   (a) you reject a null hypothesis that is true
   (b) you don't reject a null hypothesis that is true
   (c) you reject a null hypothesis that is false
   (d) you don't reject a null hypothesis that is false

3. Which of the following is an appropriate null hypothesis?
   (a) The difference between the means of two populations is equal to 0.
   (b) The difference between the means of two populations is not equal to 0.
   (c) The difference between the means of two populations is less than 0.
   (d) The difference between the means of two populations is greater than 0.

4. Which of the following is not an appropriate alternative hypothesis?
   (a) The difference between the means of two populations is equal to 0.
   (b) The difference between the means of two populations is not equal to 0.
   (c) The difference between the means of two populations is less than 0.
   (d) The difference between the means of two populations is greater than 0.

5. The power of a test is the probability of:
   (a) rejecting a null hypothesis that is true
   (b) not rejecting a null hypothesis that is true
   (c) rejecting a null hypothesis that is false
   (d) not rejecting a null hypothesis that is false

6. If the *p*-value is less than α in a two-tail test:
   (a) The null hypothesis should not be rejected.
   (b) The null hypothesis should be rejected.
   (c) A one-tail test should be used.
   (d) No conclusion can be reached.

7. A test of hypothesis has a type I error probability (α) of 0.01. Therefore:
   (a) If the null hypothesis is true, you don't reject it 1% of the time.
   (b) If the null hypothesis is true, you reject it 1% of the time.
   (c) If the null hypothesis is false, you don't reject it 1% of the time.
   (d) If the null hypothesis is false, you reject it 1% of the time.

8. Which of the following statements is not true about the level of significance in a hypothesis test?
   (a) The larger the level of significance, the more likely you are to reject the null hypothesis.
   (b) The level of significance is the maximum risk you are willing to accept in making a type I error.
   (c) The significance level is also called the α level.
   (d) The significance level is another name for a type II error.

9. If you reject the null hypothesis when it is false, then you have committed:
   (a) a type II error
   (b) a type I error
   (c) no error
   (d) a type I and type II error

10. The probability of a type _______ error is also called "the level of significance."

11. The probability of a type I error is represented by the symbol _______.

12. The value that separates a rejection region from a non-rejection region is called the _______.

13. Which of the following is an appropriate null hypothesis?
    (a) The mean of a population is equal to 100.
    (b) The mean of a sample is equal to 50.
    (c) The mean of a population is greater than 100.
    (d) All of the above.

14. Which of the following is an appropriate alternative hypothesis?
    (a) The mean of a population is equal to 100.
    (b) The mean of a sample is equal to 50.
    (c) The mean of a population is greater than 100.
    (d) All of the above.

Answer True or False:

15. For a given level of significance, if the sample size is increased, the power of the test will increase.

16. For a given level of significance, if the sample size is increased, the probability of committing a type I error will increase.

17. The statement of the null hypothesis always contains an equality.

18. The larger the *p*-value, the more likely you are to reject the null hypothesis.

19. The statement of the alternative hypothesis always contains an equality.
20. The smaller the *p*-value, the more likely you are to reject the null hypothesis.

Answers to Test Yourself

1. d
2. a
3. a
4. a
5. c
6. b
7. b
8. d
9. c
10. I
11. α
12. critical value
13. a
14. c
15. True
16. False
17. True
18. False
19. False
20. True

References

1. Berenson, M. L., D. M. Levine, and K. A. Szabat. *Basic Business Statistics: Concepts and Applications, Thirteenth Edition*. Upper Saddle River, NJ: Pearson Education, 2015.
2. Levine, D. M., D. Stephan, and K. A. Szabat. *Statistics for Managers Using Microsoft Excel, Seventh Edition*. Upper Saddle River, NJ: Pearson Education, 2014.

CHAPTER 8

Hypothesis Testing: Z and *t* Tests

In Chapter 7, you learned the fundamentals of hypothesis testing. This chapter discusses hypothesis tests that involve two groups, more formally known as two-sample tests. You will learn to use:

- The hypothesis test that examines the differences between the proportions of two groups
- The hypothesis test that examines the differences between the means of two groups

You will also learn how to evaluate the statistical assumptions about your variables that need to be true in order to use these tests and what to do if the assumptions do not hold.

8.1 Testing for the Difference Between Two Proportions

CONCEPT Hypothesis test that analyzes differences between two groups by examining the differences in sample proportions of the two groups.

INTERPRETATION The sample proportion for each group is the number of successes in the group sample divided by the group's sample size. Sample proportions for both groups are needed as the test statistic is based on the difference in the sample proportion of the two groups. With a sufficient sample size in each group, the sampling distribution of the difference between the two proportions approximately follows a normal distribution (see Section 5.3).

WORKED-OUT PROBLEM 1 Businesses use a method called *A/B testing* to test different web page designs to see if one design is more effective than another. For one company, designers were interested in the effect of modifying the call-to-action button on the home page. Every visitor to the company's home page was randomly shown either the original call-to-action button (the control) or the new variation. Designers measured success by the download rate: the number of people who downloaded the file divided by the number of people who saw that particular call-to-action button. Results of the experiment yielded the following:

		Call-to-Action Button		
		Original	**New**	**Total**
Download	**Yes**	351	451	802
	No	3,291	3,105	6,396
	Total	3,642	3,556	7,198

Of the 3,642 people who used the original call-to-action button, 351 downloaded the file, for a proportion of 0.0964. Of the 3,556 people who used the new call-to-action button, 451 downloaded the file, for a proportion of 0.1268.

Because the number of downloads triggered by the original and new call-to-action buttons is large (351 and 451) and the number of nondownloads from the original and new call-to-action buttons is also large (3,291 and 3,105), the sampling distribution for the difference between the two proportions is approximately normally distributed. The null and alternative hypotheses are as follows:

H_0: $\pi_1 = \pi_2$ (No difference in the proportion of downloads triggered by the original and new call-to-action buttons)

H_1: $\pi_1 \neq \pi_2$ (There is a difference in the proportion of downloads triggered by the original and new call-to-action buttons)

Spreadsheet results for this study are as follows:

	A	B
1	**Z Test for Differences in Two Proportions**	
2		
3	**Data**	
4	**Hypothesized Difference**	**0**
5	**Level of Significance**	**0.05**
6	**Group 1**	
7	**Number of Successes**	**351**
8	**Sample Size**	**3642**
9	**Group 2**	
10	**Number of Successes**	**451**
11	**Sample Size**	**3556**
12		
13	Intermediate Calculations	
14	Group 1 Proportion	0.0964
15	Group 2 Proportion	0.1268
16	Difference in Two Proportions	-0.0305
17	Average Proportion	0.1114
18	***Z* Test Statistic**	**-4.1052**
19		
20	**Two-Tail Test**	
21	**Lower Critical Value**	**-1.9600**
22	**Upper Critical Value**	**1.9600**
23	***p*-Value**	**0.0000**
24	**Reject the null hypothesis**	

Using the critical value approach with a level of significance of 0.05, the lower tail area is 0.025, and the upper tail area is 0.025. Using the cumulative normal distribution table (Table C.1), the lower critical value of 0.025 corresponds to a Z value of –1.96, and an upper critical value of 0.025 (cumulative area of 0.975) corresponds to a Z value of +1.96, as shown in the following diagram:

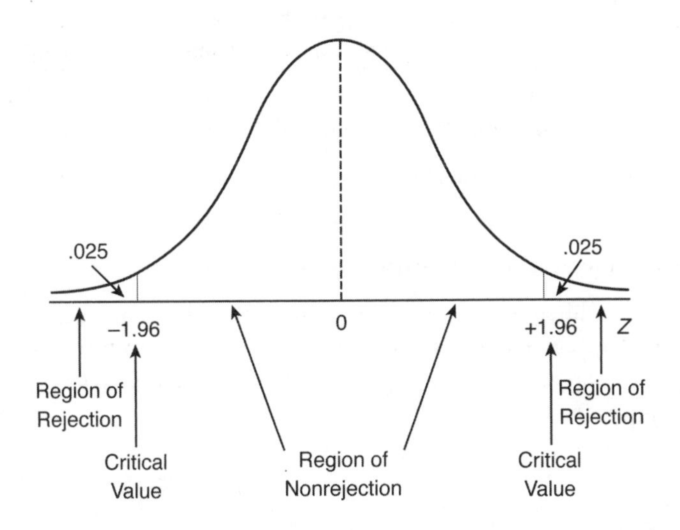

Given these rejection regions, you will reject H_0 if $Z < -1.96$ or if $Z > +1.96$; otherwise, you will not reject H_0. Spreadsheet results (see page 151) show that the Z test statistic is –4.1052. Because $Z = -4.1052$ is less than the lower critical value of –1.96, you reject the null hypothesis. You conclude that evidence exists of a difference in the proportion of downloads triggered by the original and new call-to-action buttons.

WORKED-OUT PROBLEM 2 You decide to use the *p*-value approach to hypothesis testing. Calculations done in a spreadsheet (see the figures on page 151) determine that the *p*-value is 0.0000. This means that the probability of obtaining a Z value less than –4.1052 or greater than +4.1052 is virtually zero (0.0000). Because the *p*-value is less than the level of significance $\alpha = 0.05$, you reject the null hypothesis. You conclude that evidence exists of a difference in the proportion of downloads triggered by the original and new call-to-action button. The new call-to-action button is more likely to result in downloads of the file.

WORKED-OUT PROBLEM 3 You need to analyze the results of a famous health care experiment that investigated the effectiveness of aspirin in the reduction of the incidence of heart attacks, using a level of significance of $\alpha = 0.05$. In this experiment, 22,071 male U.S. physicians were randomly assigned to either a group that was given one 325 mg buffered aspirin tablet every other day or a group that was given a placebo (a pill that contained no active ingredients). Of 11,037 physicians taking aspirin, 104 suffered heart attacks during the five-year period of the study. Of 11,034 physicians who were assigned to a group that took a placebo every other day, 189 suffered heart attacks during the five-year period of the study. You summarize these results as follows:

Results Classified by Whether the Physician Took Aspirin

		Study Group		
		Aspirin	**Placebo**	**Totals**
Results	**Heart attack**	104	189	293
	No heart attack	10,933	10,845	21,778
	Totals	11,037	11,034	22,071

You establish the null and alternative hypotheses as

H_0: $\pi_1 = \pi_2$ (No difference exists in the proportion of heart attacks between the group that was given aspirin and the group that was given the placebo.)

H_1: $\pi_1 \neq \pi_2$ (A difference exists in the proportion of heart attacks between the two groups.)

As defined by the null hypothesis, the number of successes in this problem is the number of heart attacks. This illustrates that the number of successes can represent a negative real-world outcome.

Spreadsheet results for the health care experiment data are as follows:

	A	B
1	**Z Test for Differences in Two Proportions**	
2		
3	**Data**	
4	**Hypothesized Difference**	**0**
5	**Level of Significance**	**0.05**
6	**Group 1**	
7	**Number of Successes**	**104**
8	**Sample Size**	**11037**
9	**Group 2**	
10	**Number of Successes**	**189**
11	**Sample Size**	**11034**
12		
13	Intermediate Calculations	
14	Group 1 Proportion	0.0094
15	Group 2 Proportion	0.0171
16	Difference in Two Proportions	-0.0077
17	Average Proportion	0.0133
18	**Z Test Statistic**	**-5.0014**
19		
20	**Two-Tail Test**	
21	**Lower Critical Value**	**-1.9600**
22	**Upper Critical Value**	**1.9600**
23	***p*-Value**	**0.0000**
24	**Reject the null hypothesis**	

These results show that the *p*-value is 0 and the value of the test statistic is $Z = -5.00$. This means that the chance of obtaining a Z value less than –5.00 is virtually zero. Because the *p*-value (0) is less than the level of significance (0.05), you reject the null hypothesis and accept the alternative hypothesis that a difference exists in the proportion of heart attacks between the two groups. (Using the critical value approach, because $Z = -5.00$ is less than the lower critical value of –1.96 at the 0.05 level of significance, you can reject the null hypothesis.)

You conclude that evidence exists of a difference in the proportion of doctors who have had heart attacks between those who took the aspirin and those who did not take the aspirin. The study group who took the aspirin had a significantly lower proportion of heart attacks over the study period.

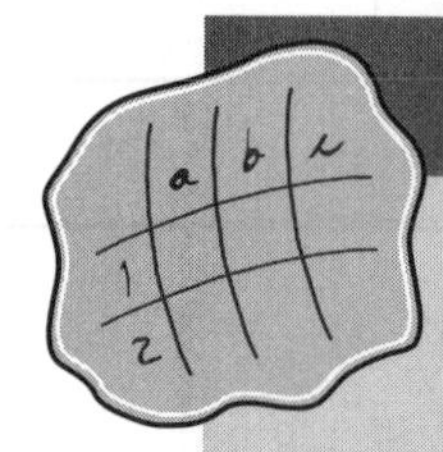

spreadsheet solution

Z Test for the Difference in Two Proportions

Chapter 8 Z Two Proportions contains the Z test for the difference in two proportions for WORKED-OUT PROBLEM 1, shown on page 150. Experiment with this spreadsheet by changing the hypothesized difference and level of significance in cells B4 and B5, and the number of successes and sample size for each group in cells B7, B8, B10, and B11.

Best Practices

Use the **NORM.S.INV(*P*<*X*)** function to calculate a critical value of the normal distribution, where *P*<*X* is the area under the curve that is less than *X*.

Use the **NORM.S.DIST(Z *value*, TRUE)** function to calculate the cumulative normal probability of less than the Z test statistic.

How Tos

For the two-tail test, compute the *p*-value by multiplying 2 by the expression (1 minus the absolute value of the NORM.S.DIST value).

interested in math?

WORKED-OUT PROBLEMS 1–3 use the Z test for the difference between two proportions. You need the subscripted symbols for the number of successes, X, the sample sizes, n_1 and n_2, sample proportions, p_1 and p_2, and population proportions, π_1 and π_2, as well as the symbol for the pooled estimate of the population proportion, $\bar{p}$, to calculate the Z test statistic.

To write the Z test statistic equation, you first define the symbols for the pooled estimate of the population proportion and the sample proportions for the two groups:

$$\bar{p} = \frac{X_1 + X_2}{n_1 + n_2} \quad p_1 = \frac{X_1}{n_1} \quad p_2 = \frac{X_2}{n_2}$$

Next, you use $\bar{p}$, p_1, and p_2 along with the symbols for the sample sizes and population proportion to form the equation for the Z test for the difference between two proportions:

$$Z = \frac{(p_1 - p_2) - (\pi_1 - \pi_2)}{\sqrt{\bar{p}(1-\bar{p})\left(\frac{1}{n_1} + \frac{1}{n_2}\right)}}$$

As an example, the calculations for determining the Z test statistic for WORKED-OUT PROBLEM 1 concerning the web design are as follows:

$$p_1 = \frac{X_1}{n_1} = \frac{351}{3{,}642} = 0.0964 \quad p_2 = \frac{X_2}{n_2} = \frac{451}{3{,}556} = 0.1268$$

and

$$\bar{p} = \frac{X_1 + X_2}{n_1 + n_2} = \frac{351 + 451}{3{,}642 + 3{,}556} = \frac{802}{7{,}198} = 0.1114$$

so that

$$\begin{aligned} Z_{STAT} &= \frac{(0.0964 - 0.1268) - (0)}{\sqrt{0.1114(1 - 0.1114)\left(\frac{1}{3{,}642} + \frac{1}{3{,}556}\right)}} \\ &= \frac{-0.0304}{\sqrt{(0.09899)(0.0005557)}} \\ &= \frac{-0.0304}{\sqrt{0.000055}} \\ &= \frac{-0.0304}{0.00742} \\ &= -4.10 \end{aligned}$$

With $\alpha = 0.05$, you reject H_0 if $Z < -1.96$ or if $Z > +1.96$; otherwise, do not reject H_0. Because $Z = -4.10$ is less than the lower critical value of -1.96, you reject the null hypothesis.

8.2 Testing for the Difference Between the Means of Two Independent Groups

CONCEPT Hypothesis test that analyzes differences between two groups by determining whether a significant difference exists in the population means (a numerical parameter) of two populations or groups.

INTERPRETATION Statisticians distinguish between using two independent groups and using two related groups when performing this type of hypothesis test. With related groups, the observations are either matched according to a relevant characteristic or repeated measurements of the same items are taken. For studies involving two independent groups, the most common test of hypothesis used is the pooled-variance *t* test.

Pooled-Variance *t* Test

CONCEPT The hypothesis test for the difference between the population means of two independent groups that combines or "pools" the sample variance of each group into one estimate of the variance common in the two groups.

INTERPRETATION For this test, the test statistic is based on the difference in the sample means of the two groups, and the sampling distribution for the difference in the two sample means approximately follows the *t* distribution.

In a pooled variance *t* test, the null hypothesis of no difference in the means of two independent populations is

H_0: $\mu_1 = \mu_2$ (The two population means are equal.)

and the alternative hypothesis is

H_1: $\mu_1 \neq \mu_2$ (The two population means are not equal.)

WORKED-OUT PROBLEM 4 You want to determine whether the cost of a restaurant meal in a major city differs from the cost of a similar meal in the suburbs outside the city. You collect data about the cost of a meal per person from a sample of 50 city restaurants and 50 suburban restaurants as follows:

Restaurants

City Cost Data

25 26 27 29 32 32 33 33 34 35 35 36 37 39 41 42 42 43 43 43 44 44 44 44 45
48 50 50 50 50 51 53 54 55 56 57 57 60 61 61 65 66 67 68 74 74 76 77 77 80

Suburban Cost Data

26 27 28 29 31 33 34 34 34 34 34 34 35 36 37 37 37 38 39 39 39 40 41 41 43
44 44 44 46 47 47 48 48 49 50 51 51 51 51 52 52 54 56 59 60 60 67 68 70 71

The Excel results for the cost of the restaurant meals are as follows:

	A	B
1	**Pooled-Variance t Test for Differences in Two Means**	
2	(assumes equal population variances)	
3	**Data**	
4	**Hypothesized Difference**	**0**
5	**Level of Significance**	**0.05**
6	**Population 1 Sample**	
7	**Sample Size**	**50**
8	**Sample Mean**	**49.3**
9	**Sample Standard Deviation**	**14.9151**
10	**Population 2 Sample**	
11	**Sample Size**	**50**
12	**Sample Mean**	**44.4**
13	**Sample Standard Deviation**	**11.3785**
14		
15	Intermediate Calculations	
16	Population 1 Sample Degrees of Freedom	49
17	Population 2 Sample Degrees of Freedom	49
18	Total Degrees of Freedom	98
19	Pooled Variance	175.9643
20	Standard Error	2.6530
21	Difference in Sample Means	4.9
22	***t* Test Statistic**	**1.8469**
23		
24	**Two-Tail Test**	
25	**Lower Critical Value**	**-1.9845**
26	**Upper Critical Value**	**1.9845**
27	***p-Value***	**0.0678**
28	**Do not reject the null hypothesis**	

These results show that the t statistic is 1.8469 and the p-value is 0.0678. Because $t = 1.8469 < 1.9845$ or because the p-value, 0.0678, is greater than $\alpha = 0.05$, you do not reject the null hypothesis. You conclude that you have insufficient evidence of a significant difference in the cost of a restaurant meal in the city (sample mean is $49.30) and a restaurant meal in the suburbs outside the city (sample mean of $44.40).

WORKED-OUT PROBLEM 5 You want to determine at a level of significance of $\alpha = 0.05$ whether the mean payment made by online customers of a website differs according to two methods of payment. You obtained the following statistics based on a random sample of 50 transactions.

	Method 1	Method 2
Sample Size	22.0	28.0
Sample Mean	30.37	23.17
Sample Standard Deviation	12.006	7.098

(*continues on page* 159)

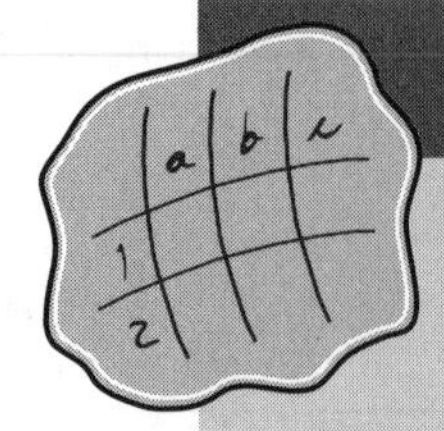

spreadsheet solution

Pooled-Variance *t* Test for the Differences in Two Means with Unsummarized Data

Chapter 8 Pooled-Variance *t* with Unsummarized Data contains the WORKED-OUT PROBLEM 4 pooled-variance *t* test for the difference in two means, shown above. Experiment with this spreadsheet by changing the hypothesized difference and level of significance in cells B4 and B5 and/or entering new, unsummarized data in columns E and F, replacing the city and suburban restaurant meal cost data.

Best Practices

Use the **T.INV.2T(*level of confidence, degrees of freedom)*** function to calculate the upper critical value of the *t* distribution. Precede the same function with a minus sign to calculate the lower critical value of the *t* distribution.

Use the **T.DIST.2T(*absolute value of the t test statistic), total degrees of freedom)*** function to calculate a probability associated with the *t* distribution.

How Tos

To calculate the absolute value of the *t* test statistic, use the **ABS(*t* test statistic)** function.

Tip ATT3 in Appendix E describes using the Analysis ToolPak as a second way to perform a pooled-variance *t* test for the difference in two means using unsummarized data.

Spreadsheet results for this study are as follows:

	A	B
1	**Pooled-Variance *t* Test for Differences in Two Means**	
2	(assumes equal population variances)	
3	**Data**	
4	**Hypothesized Difference**	**0**
5	**Level of Significance**	**0.05**
6	**Population 1 Sample**	
7	**Sample Size**	**22**
8	**Sample Mean**	**30.37**
9	**Sample Standard Deviation**	**12.0060**
10	**Population 2 Sample**	
11	**Sample Size**	**28**
12	**Sample Mean**	**23.17**
13	**Sample Standard Deviation**	**7.0980**
14		
15	Intermediate Calculations	
16	Population 1 Sample Degrees of Freedom	21
17	Population 2 Sample Degrees of Freedom	27
18	Total Degrees of Freedom	48
19	Pooled Variance	91.4027
20	Standard Error	2.7238
21	Difference in Sample Means	7.2
22	***t* Test Statistic**	**2.6434**
23		
24	**Two-Tail Test**	
25	**Lower Critical Value**	**-2.0106**
26	**Upper Critical Value**	**2.0106**
27	***p-Value***	**0.0111**
28	**Reject the null hypothesis**	

The results show that the t statistic is 2.64 and the p-value is 0.0111. Because the p-value is less than $\alpha = 0.05$, you reject the null hypothesis. (Using the critical value approach, $t = 2.6434 > 2.01$, leading you to the same decision.) You conclude that the chance of obtaining a t value greater than 2.64 is very small (0.0111) and therefore assert that the mean payment amount is higher for method 1 (sample mean of \$30.37) than for method 2 (sample mean of \$23.17).

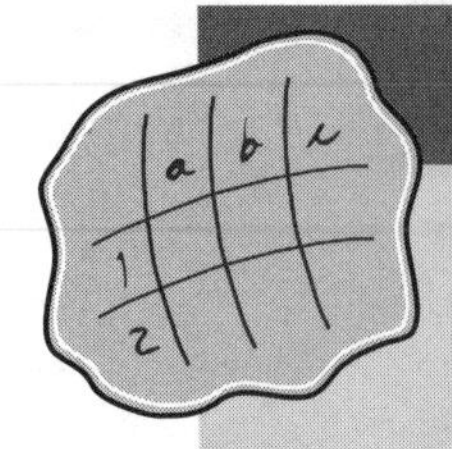

spreadsheet solution

Pooled-Variance *t* Test for the Differences in Two Means with Sample Statistics

Chapter 8 Pooled-Variance T with Sample Statistics contains the WORKED-OUT PROBLEM 5 pooled-variance *t* test for the difference in two means, shown on page 159. Experiment with this spreadsheet by changing the hypothesized difference and level of significance in cells B4 and B5, and the sample size, sample mean, and sample standard deviation for each group in cells B7 through B9 and B11 through B13.

Best Practices

This spreadsheet uses the functions that are described in the Spreadsheet Solution for using unsummarized data on page 158.

WORKED-OUT PROBLEMS 4 and 5 use the pooled variance *t* test for the difference between the population means of two independent groups. You need the subscripted symbols for the sample means, $\overline{X}_1$ and $\overline{X}_2$, the sample sizes for each of the two groups, n_1 and n_2, and the population means, μ_1 and μ_2 along with the symbol for the pooled estimate of the variance, S_p^2, to calculate the *t* test statistic.

To write the equation for the *t* test statistic, you first define the symbols for the equation for the pooled estimate of the population variance:

$$S_p^2 = \frac{(n_1 - 1)S_1^2 + (n_2 - 1)S_2^2}{(n_1 - 1) + (n_2 - 1)}$$

You next use S_p^2 that you just defined and the symbols for the sample means, the population means, and the sample sizes to form the equation for the pooled-variance t test for the difference between two means:

$$t = \frac{(\bar{X}_1 - \bar{X}_2) - (\mu_1 - \mu_2)}{\sqrt{S_p^2\left(\frac{1}{n_1} + \frac{1}{n_2}\right)}}$$

The calculated test statistic t follows a t distribution with $n_1 + n_2 - 2$ degrees of freedom.

As an example, the calculations for determining the t test statistic for the restaurant cost data with $\alpha = 0.05$, are as follows:

$$S_p^2 = \frac{(n_1 - 1)S_1^2 + (n_2 - 1)S_2^2}{(n_1 - 1) + (n_2 - 1)}$$

$$= \frac{49(222.4592) + 49(129.4694)}{49 + 49} = 175.9643$$

Using 175.9643 as the value for S_p^2 in the original equation

$$t = \frac{(\bar{X}_1 - \bar{X}_2) - (\mu_1 - \mu_2)}{\sqrt{S_p^2\left(\frac{1}{n_1} + \frac{1}{n_2}\right)}}$$

produces

$$t = \frac{(49.30 - 44.40) - 0}{\sqrt{175.9643\left(\frac{1}{50} + \frac{1}{50}\right)}}$$

$$t = \frac{49.30 - 44.40}{\sqrt{175.9643(0.04)}}$$

$$= \frac{4.90}{\sqrt{7.0386}} = +1.8469$$

Using the $\alpha = 0.05$ level of significance, with $50 + 50 - 2 = 98$ degrees of freedom, the critical value of t is 1.9845 (0.025 in the upper tail of the t distribution). Because $t = +1.8469 < 1.9845$, you don't reject H_0.

Pooled-Variance *t* Test Assumptions

In testing for the difference between the means, you assume that the two populations from which the two independent samples have been selected are normally distributed with equal variances. When the two populations *do* have equal variances, the pooled-variance t test is valid even if there is a moderate departure from normality, as long as the sample sizes are large.

You can check the assumption of normality by preparing a side-by-side box-and-whisker plot for the two samples. Such a plot was shown for the meal cost at city and suburban restaurants in Chapter 3 on page 55. These box-and-whisker plots have some right-skewness because the tail on the right is longer than the one on the left. However, given the relatively large sample size in each of the two groups, you can conclude that any departure from the normality assumption will not seriously affect the validity of the t test. If the data in each group cannot be assumed to be from normally distributed populations, you can use a nonparametric procedure, such as the Wilcoxon rank sum test (see references 1 and 2) that does not depend on the assumption of normality in the two populations.

The pooled-variance t test also assumes that the population variances are equal. If this assumption cannot be made, you cannot use the pooled-variance t test. In such cases, you can use the separate-variance t test (see references 1 and 2).

8.3 The Paired *t* Test

CONCEPT A hypothesis test for the difference between two groups for situations in which the data from the two groups is *related and not independent*. In this test, the variable of interest is the difference between related pairs of values in the two groups, rather than the paired values themselves.

INTERPRETATION There are two situations in which the values from two groups will be related and not independent.

In the first case, a researcher has paired, or matched, the values under study according to some other variable. For example, in testing whether a new drug treatment lowers blood pressure, a sample of patients could be paired according to their blood pressure at the beginning of the study. For example, if there were two patients in the study who each had a diastolic blood pressure of 140, one would be randomly assigned to the group that will take the new drug and the other to the group that will not take the new drug. Assigning patients in this manner means that the researcher will not have to be concerned about differences in the initial blood pressures of the patients that form the two groups. This, in turn, means that test results will better reflect the effect of the new drug being tested.

In the second case, a researcher obtains two sets of measurements from the same items or individuals. This approach is based on the theory that the same items or individuals will behave alike if treated alike. This, in turn, allows the researcher to assert that any differences between two sets of measurements are due to what is under study. For example, when performing an experiment on the effect of a diet drug, the researcher could take one measurement from each participant just prior to starting the drug and one just after the end of a specified time period.

In both cases, the variable of interest can be stated algebraically as follows:

Difference (D) = Related value in sample 1 – Related value in sample 2

With related groups and a numerical variable of interest, the null hypothesis is that no difference exists in the population means of the two related groups, and the alternative hypothesis is that there is a difference in the population means of the two related groups. Using the symbol μ_D to represent the difference between the population means, the null and alternative hypotheses can be expressed as follows:

$$H_0: \mu_D = 0$$

and

$$H_1: \mu_D \neq 0$$

To decide whether to reject the null hypothesis, you use a paired t test.

WORKED-OUT PROBLEM 6 You want to determine whether there is a difference between the rating of TV service and Internet Service for various providers. Data were available from 13 different providers, as follows:

Telecom

Provider	TV	Internet	Difference
Verizon FIOS	73	74	−1
WOW	74	76	−2
Bright House Networks	68	70	−2
AT&T U-verse	68	68	0
Cox	64	68	−4
SuddenLink	65	70	−5
Cablevision/Optimum	63	67	−4
RCN	65	71	−6
Comcast/Xfinity	59	62	−3
TimeWarner	58	63	−5
Charter	59	64	−5
Mediacom	54	58	−4
Wave/Astound	72	74	−2
Cable One	63	68	−5

Source: Data extracted from "Ratings: TV, Phone, and Internet Services," *Consumer Reports*, May 2014, pp. 28–29.

Because the two sets of ratings are from the same providers, the two sets of measurements are related and only the differences between the TV and Internet service for the providers are tested.

Spreadsheet results are as follows:

	A	B
1	**Paired *t* Test**	
2		
3	**Data**	
4	**Hypothesized Mean Diff.**	**0**
5	**Level of Significance**	**0.05**
6		
7	Intermediate Calculations	
8	Sample Size	14
9	Mean Difference	-3.4286
10	Degrees of Freedom	13
11	Sample Standard Deviation	1.7852
12	Standard Error	0.4771
13	***t* Test Statistic**	**-7.1862**
14		
15	**Two-Tailed Test**	
16	**Lower Critical Value**	**-2.1604**
17	**Upper Critical Value**	**2.1604**
18	***p*-Value**	**0.0000**
19	**Reject the null hypothesis**	

The results show that the *t* statistic is –7.1862 and the *p*-value is 0.0000. Because the *p*-value is 0.0000 is less than $\alpha = 0.05$ (or because $t = -7.1862 < -2.1604$), you reject the null hypothesis. This means that the chance of obtaining a *t* value less than –7.1862 is virtually zero and you conclude that a difference exists between the TV and Internet service ratings. The TV ratings are lower than Internet service ratings.

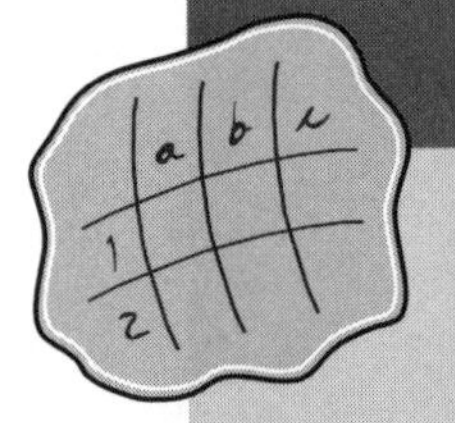

spreadsheet solution

Paired *t* Test

Chapter 8 Paired *t* contains the WORKED-OUT PROBLEM 6 paired *t* test for the difference in two means, shown on this page. Experiment with this spreadsheet by changing the hypothesized mean difference and level of significance in cells B4 and B5 and/or entering new unsummarized data in columns E through G, replacing the telecommunications ratings data.

Best Practices

Use the T.INV.2T(*level of confidence, degrees of freedom*) function to calculate the upper critical value of the *t* distribution. Precede the same function with a minus sign to calculate the lower critical value of the *t* distribution.

Use the T.DIST.2T(*absolute value of the t test statistic), total degrees of freedom)* function to calculate a probability associated with the *t* distribution.

How Tos

To help calculate the sample standard deviation, use the **DEVSQ(*column of differences*)** function to compute the sum of the squares of the differences between each set of paired values and the mean difference.

Tip ATT4 in Appendix E describes using the Analysis ToolPak as a second way to perform a paired t test for the difference in two means.

Tip ADV5 in Appendix E explains how to modify the spreadsheet for data sets that have fewer than or more than 15 rows of values.

equation blackboard (optional)

interested in math?

WORKED-OUT PROBLEM 6 uses the equation for the paired t test. You need the symbols for the sample size, n, the difference between the population means, μ_D, the sample standard deviation, S_D, and the subscripted symbol for the differences in the paired values, D_i, all previously introduced, and the symbol for the mean difference, $\bar{D}$, to calculate the t test statistic.

To write the t test statistic equation, you first define the symbols for the equation for the mean difference, $\bar{D}$:

$$\bar{D} = \frac{\sum_{i=1}^{n} D_i}{n}$$

You next use $\bar{D}$, the symbols for the sample size, and the differences in the paired values to form the equation for the sample standard deviation, S_D:

$$S_D = \sqrt{\frac{\sum_{i=1}^{n} (D_i - \bar{D})^2}{n-1}}$$

Finally, you assemble $\bar{D}$ and S_D and the remaining symbols to form the equation for the paired t test for the difference between two means:

$$t = \frac{\bar{D} - \mu_D}{\frac{S_D}{\sqrt{n}}}$$

(*continues*)

The test statistic t follows a t distribution with $n - 1$ degrees of freedom.

As an example, the calculations for determining the t test statistic for WORKED-OUT PROBLEM 6 concerning the difference between the TV and Internet service ratings measurements with $\alpha = 0.05$, are as follows:

$$\bar{D} = \frac{\sum_{i=1}^{n} D_i}{n}$$

$$= \frac{-48}{14} = -3.4286$$

This makes $S_D = 1.7852$ (calculation not shown). Substituting these values results in

$$t = \frac{\bar{D} - \mu_D}{\frac{S_D}{\sqrt{n}}} = \frac{-3.4286 - 0}{\frac{1.7852}{\sqrt{14}}} = -7.1862$$

Using the $\alpha = 0.05$ level of significance, with $14 - 1 = 13$ degrees of freedom, the critical value of t is -2.1604 (0.025 in the lower tail of the t distribution). Because $t = -7.1862 < -2.1604$, you reject the null hypothesis H_0.

WORKED-OUT PROBLEM 7 You seek to determine, using a level of significance of $\alpha = 0.05$, whether differences exist in monthly sales between the new package design and the old package design of a laundry stain remover. The new package was test marketed over a period of one month in a sample of supermarkets in a particular city. A random sample of ten pairs of supermarkets was matched according to weekly sales volume and a set of demographic characteristics. The data collected for this study are as follows:

Monthly Sales of Laundry Stain Remover

Supermarket

Pair	New Package	Old Package	Difference
1	458	437	21
2	519	488	31
3	394	409	−15
4	632	587	45
5	768	753	15
6	348	400	−52
7	572	508	64
8	704	695	9
9	527	496	31
10	584	513	71

Because the ten pairs of supermarkets were matched, you use the paired *t* test. The spreadsheet results for this study are as follows:

	A	B
1	**Paired *t* Test**	
2		
3	**Data**	
4	**Hypothesized Mean Diff.**	**0**
5	**Level of Significance**	**0.05**
6		
7	Intermediate Calculations	
8	Sample Size	10
9	Mean Difference	22.0000
10	Degrees of Freedom	9
11	Sample Standard Deviation	36.3929
12	Standard Error	11.5085
13	***t* Test Statistic**	**1.9116**
14		
15	**Two-Tailed Test**	
16	**Lower Critical Value**	**-2.2622**
17	**Upper Critical Value**	**2.2622**
18	***p*-Value**	**0.0882**
19	**Do not reject the null hypothesis**	

These results show that the *t* statistic is 1.91 and the *p*-value is 0.0882. This means that the chance of obtaining a *t* value greater than 1.91 or less than –1.91 is 0.0882 or 8.82%. Because the *p*-value is 0.0882 is greater than $\alpha = 0.05$, you do not reject the null hypothesis. (Using the critical value approach, $t = 1.91 < 2.2622$, and you reach the same decision.) You can conclude that insufficient evidence exists of a difference between the new and old package design.

WORKED-OUT PROBLEM 8 Before the test marketing experiment, you might have wanted to determine whether the new package design produced *more* sales than the old package design and not just a difference in sales. Such a situation would be a good application of a one-tail test in which the alternative hypothesis is H_1: $\mu_D > 0$.

In such a case, you would use the one-tail *p*-value (or one-tail critical value). For the test market sales data, the one-tail *p*-value is 0.0441 (and the one-tail critical value is 1.8331). Because the *p*-value is less than $\alpha = 0.05$ (or because using the critical value approach, $t = 1.911$ is greater than 1.8331), you reject the null hypothesis and conclude that the mean sales from the new package design were higher than the mean sales for the old package design—a different conclusion than you made using the two-tail test.

Important Equations

Z Test for the Difference Between Two Proportions:

$$Z = \frac{(p_1 - p_2) - (\pi_1 - \pi_2)}{\sqrt{\bar{p}(1-\bar{p})\left(\frac{1}{n_1} + \frac{1}{n_2}\right)}} \quad (8.1)$$

Pooled Variance t Test for the Difference Between the Population Means of Two Independent Groups:

$$t = \frac{(\bar{X}_1 - \bar{X}_2) - (\mu_1 - \mu_2)}{\sqrt{S_p^2\left(\frac{1}{n_1} + \frac{1}{n_2}\right)}} \quad (8.2)$$

Paired t Test for the Difference Between Two Means:

$$t = \frac{\bar{D} - \mu_D}{\frac{S_D}{\sqrt{n}}} \quad (8.3)$$

One-Minute Summary

For tests for the differences between two groups, first determine whether your data are categorical or numerical:

- If your data are categorical, use the Z test for the difference between two proportions.
- If your data are numerical, determine whether you have independent or related groups:

  If you have independent groups, use the pooled variance t test for the difference between two means.

  If you have related groups, use the paired t test.

Test Yourself

Short Answers

1. The t test for the difference between the means of two independent populations assumes that the two:
   (a) Sample sizes are equal.
   (b) Sample medians are equal.
   (c) Populations are approximately normally distributed.
   (d) All of the above.

2. In testing for differences between the means of two related populations, the null hypothesis is:
   (a) $H_0: \mu_D = 2$
   (b) $H_0: \mu_D = 0$
   (c) $H_0: \mu_D < 0$
   (d) $H_0: \mu_D > 0$

3. A researcher is curious about the effect of sleep on students' test performances. He chooses 100 students and gives each student two exams. One is given after four hours' sleep and one after eight hours' sleep. The statistical test the researcher should use is the:
   (a) Z test for the difference between two proportions
   (b) Pooled-variance t test
   (c) Paired t test

4. A statistics professor wanted to test whether the grades on a statistics test were the same for her morning class and her afternoon class. For this situation, the professor should use the:
   (a) Z test for the difference between two proportions
   (b) Pooled-variance t test
   (c) Paired t test

Answer True or False:

5. The sample size in each independent sample must be the same in order to test for differences between the means of two independent populations.

6. In testing a hypothesis about the difference between two proportions, the p-value is computed to be 0.043. The null hypothesis should be rejected if the chosen level of significance is 0.05.

7. In testing a hypothesis about the difference between two proportions, the p-value is computed to be 0.034. The null hypothesis should be rejected if the chosen level of significance is 0.01.

8. In testing a hypothesis about the difference between two proportions, the Z test statistic is computed to be 2.04. The null hypothesis should be rejected if the chosen level of significance is 0.01 and a two-tail test is used.
9. The sample size in each independent sample must be the same in order to test for differences between the proportions of two independent populations.
10. When you are sampling the same individuals and taking a measurement before treatment and after treatment, you should use the paired t test.
11. Repeated measurements from the same individuals are an example of data collected from two related populations.
12. The pooled-variance t test assumes that the population variances in the two independent groups are equal.
13. In testing a null hypothesis about the difference between two proportions, the Z test statistic is computed to be 2.04. The p-value is 0.0207.
14. You can use a pie chart to evaluate whether the assumption of normally distributed populations in the pooled-variance t test has been violated.
15. If the assumption of normally distributed populations in the pooled-variance t test has been violated, you should use an alternative procedure such as the nonparametric Wilcoxon rank sum test.

Answers to Test Yourself Short Answers

1. c
2. b
3. c
4. b
5. False
6. True
7. False
8. False
9. False
10. True
11. True
12. True
13. False
14. False
15. True

Problems

1. Are men and women equally likely to say that a major reason they use Facebook is to share with many people at once? A survey reported that 42% of men (193 out of 459 sampled) and 50% of women (250 out of 501 sampled) said that a major reason they use Facebook is to share with many people at once. (Source: "6 new facts about Facebook," bit.ly/1rCTrOO.)

At the 0.05 level of significance, is there evidence of a difference between men and women in the proportion that say that a major reason they use Facebook is to share with many people at once?

2. The owner of a restaurant that serves continental-style entrées wants to learn more about the patterns of patron demand during the Friday-to-Sunday time period. She has decided to study the demand for dessert during this period. In addition to studying whether a dessert was ordered, she will study whether a beef entrée was ordered. Data were collected from 630 customers. Of the 197 patrons who ordered a beef entrée, 74 ordered dessert. Of the 433 patrons who did not order a beef entrée, 68 ordered dessert.

   At the 0.05 level of significance, is there evidence of a difference in the proportion who order dessert based on whether they ordered a beef entrée?

3. Do people trust banks to do what is right? A survey done in the United States and Japan revealed that, of 500 respondents in the United States, 250 said that they trusted banks to do what is right; in Japan, of 200 respondents, 120 said that they trusted banks to do what is right. (Source: F. Norris, "Where Banking Crisis Raged, Trust Is Slow to Return," *The New York Times*, 26 January 2013, p. B3.)

   At the 0.05 level of significance, is there evidence of a difference between respondents in the United States and Japan who trust banks to do what is right?

4. When people make estimates, they are influenced by anchors to their estimates. A study was conducted in which students were asked to estimate the number of calories in a cheeseburger. One group was asked to do this after thinking about a calorie-laden cheesecake. A second group was asked to do this after thinking about an organic fruit salad. The mean number of calories estimated in a cheeseburger was 780 for the group that thought about the cheesecake and 1,041 for the group that thought about the organic fruit salad. (Source: "Drilling Down, Sizing Up a Cheeseburger's Caloric Heft," *The New York Times*, 4 October 2010, p. B2.)

   Suppose that the study was based on a sample of 20 people who thought about the cheesecake first and 20 people who thought about the organic fruit salad first. Also suppose that the standard deviation of the number of calories in the cheeseburger was 128 for the people who thought about the cheesecake first and 140 for the people who thought about the organic fruit salad first.

   (a) State the null and alternative hypotheses if you want to determine whether there is a difference in the mean estimated number of calories in the cheeseburger for the people who thought about the cheesecake first and for the people who thought about the organic fruit salad first.

(b) At the 0.05 level of significance, is there evidence of a difference in the mean estimated number of calories in the cheeseburger for the people who thought about the cheesecake first and those who thought about the organic fruit salad first?

5. You would like to determine at a level of significance of $\alpha = 0.05$, whether the mean surface hardness of steel intaglio printing plates prepared using a new treatment differs from the mean hardness of plates that are untreated. The following results are from an experiment in which 40 steel plates, 20 treated and 20 untreated, were tested for surface hardness.

Intaglio

Surface Hardness of 20 Untreated Steel Plates and 20 Treated Steel Plates

Untreated		Treated	
164.368	177.135	158.239	150.226
159.018	163.903	138.216	155.620
153.871	167.802	168.006	151.233
165.096	160.818	149.654	158.653
157.184	167.433	145.456	151.204
154.496	163.538	168.178	150.869
160.920	164.525	154.321	161.657
164.917	171.230	162.763	157.016
169.091	174.964	161.020	156.670
175.276	166.311	167.706	147.920

At the 0.05 level of significance, is there evidence of a difference in the mean surface hardness of steel intaglio printing plates that are untreated and those prepared using a new treatment?

6. Telephone line problems are upsetting to both customers and the telephone provider. The following table presents two samples of 20 times to clear a problem (in minutes) reported to two different central offices of a telephone provider.

Phone

Central Office I Time to Clear Problems (minutes)

1.48	1.75	0.78	2.85	0.52	1.60	4.15	3.97	1.48	3.10
1.02	0.53	0.93	1.60	0.80	1.05	6.32	3.93	5.45	0.97

Central Office II Time to Clear Problems (minutes)

7.55	3.75	0.10	1.10	0.60	0.52	3.30	2.10	0.58	4.02
3.75	0.65	1.92	0.60	1.53	4.23	0.08	1.48	1.65	0.72

Is there evidence of a difference in the mean waiting time between the two offices? (Use $\alpha = 0.05$.)

Target Walmart

7. A Florida newspaper compared the prices of the same 33 grocery items at local Target and Walmart stores and found the prices that are stored in the **Target Walmart** file. (Data extracted from "Supermarket Showdown," *The Palm Beach Post*, 13 February 2011, p. 1F, 2F.)

   At the 0.05 level of significance, is there evidence of a difference in the mean price between Target and Walmart?

8. Multiple myeloma, or blood plasma cancer, is characterized by increased blood vessel formulation (angiogenesis) in the bone marrow that is a prognostic factor in survival. One treatment approach used for multiple myeloma is stem cell transplantation with the patient's own stem cells. The following data represent the bone marrow microvessel density for patients who had a complete response to the stem cell transplant, as measured by blood and urine tests. The measurements were taken immediately prior to the stem cell transplant and at the time of the complete response:

Myeloma

Patient	Before	After
1	158	284
2	189	214
3	202	101
4	353	227
5	416	290
6	426	176
7	441	290

Source: Extracted from S. V. Rajkumar, R. Fonseca, T. E. Witzig, M. A. Gertz, and P. R. Greipp, Bone Marrow Angiogenesis in Patients Achieving Complete Response After Stem Cell Transplantation for Multiple Myeloma, *Leukemia*, 1999, 13, pp. 469–472.

   At the 0.05 level of significance, is there evidence of a difference in the mean bone marrow microvessel density before the stem cell transplant and after the stem cell transplant?

Concrete

9. The Concrete file contains data that represent the compressive strength, in thousands of pounds per square inch (psi), of 40 samples of concrete taken two and seven days after pouring. (Data extracted from O. Carrillo-Gamboa and R. F. Gunst, "Measurement-Error-Model Collinearities," *Technometrics*, 34, 1992, pp. 454–464.)

   At the 0.01 level of significance, is there evidence that the mean strength is lower at two days than at seven days?

Answers to Test Yourself Problems

1. $Z = -2.4379 < -1.96$ (or p-value $= 0.0148 < 0.05$), reject H_0. There is evidence of a difference between men and women in the proportion

who say that a major reason they use Facebook is sharing with many people at once.

2. $Z = 6.0873 > 1.96$ (or p-value $= 0.0000 < 0.05$), reject H_0. There is evidence of a difference in the proportion who order dessert based on whether they ordered a beef entrée.

3. $Z = -2.3944 < -1.96$ (or p-value $= 0.0166 < 0.05$), reject H_0. There is evidence of a difference in the proportion who trust banks to do what is right between respondents in the United States and Japan.

4. (a) H_0: $\mu_1 = \mu_2$ (The two population means are equal.) The alternative hypothesis is H_1: $\mu_1 \neq \mu_2$ (The two population means are not equal.)
   (b) Because $t = -6.1532 < -2.0244$ (or p-value $= 0.0000 < 0.05$), reject H_0. There is evidence that a difference in the mean estimated number of calories in the cheeseburger for the people who thought about the cheesecake first and those who thought about the organic fruit salad first.

5. Because $t = 4.104 > 2.0244$ (or p-value $= 0.0002 < 0.05$), reject H_0. There is evidence that a difference exists in the mean surface hardness of steel intaglio printing plates that are untreated and those prepared using a new treatment.

6. Because $t = 0.3544 < 2.0244$ (or p-value $= 0.7250 > 0.05$), do not reject H_0. There is insufficient evidence that a difference exists in the mean waiting time between the two offices.

7. Because $t = 1.7948 < 2.0369$ (or p-value $= 0.0821 > 0.05$), do not reject H_0. There is insufficient evidence of a difference in the mean price at Target and Walmart.

8. Because $t = 1.8426 < 2.4469$ (or p-value $= 0.1150 > 0.05$), do not reject H_0. There is insufficient evidence of a difference in the mean bone marrow microvessel density before the stem cell transplant and after the stem cell transplant.

9. Because $t = -9.3721 < -2.4258$ (or p-value $= 0.0000 > 0.01$), reject H_0. There is evidence that the mean strength is lower at two days than at seven days.

References

1. Berenson, M. L., D. M. Levine, and K. A. Szabat. *Basic Business Statistics: Concepts and Applications*, Thirteenth Edition. Upper Saddle River, NJ: Pearson Education, 2015.

2. Levine, D. M., D. Stephan, and K. A. Szabat. *Statistics for Managers Using Microsoft Excel*, Seventh Edition. Upper Saddle River, NJ: Pearson Education, 2014.

3. Microsoft Excel 2013. Redmond, WA: Microsoft Corporation, 2012.

CHAPTER 9

Hypothesis Testing: Chi-Square Tests and the One-Way Analysis of Variance (ANOVA)

In Chapter 8, you learned several hypothesis tests that you use to analyze differences between two groups. In this chapter, you learn about tests that you can use when you have multiple (two or more) groups.

9.1 Chi-Square Test for Two-Way Cross-Classification Tables

CONCEPT The hypothesis tests for the difference in the proportion of successes in two or more groups or a relationship between two categorical variables in a two-way cross-classification table.

INTERPRETATION Recall from Chapter 2 that a two-way cross-classification table presents the count of joint responses to two categorical variables. The categories of one variable form the rows of the table and the categories of the other variable form the columns. The chi-square test determines whether a relationship exists between the row variable and the column variable.

The null and alternative hypotheses for the two-way cross-classification table are

H_0: (There is no relationship between the row variable and the column variable.)

H_1: (There is a relationship between the row variable and the column variable.)

For the special case of a table that contains only two rows and two columns, the chi-square test becomes equivalent to the Z test for the difference between two proportions discussed in Section 8.1. The null and alternative hypotheses are restated as

H_0: $\pi_1 = \pi_2$ (No difference exists between the two proportions.)
H_1: $\pi_1 \neq \pi_2$ (A difference exists between the two proportions.)

The chi-square test compares the actual count (or frequency) in each cell, the intersection of a row and column, with the frequency that would be expected to occur if the null hypothesis were true. The expected frequency for each cell is calculated by multiplying the row total of that cell by the column total of that cell and dividing by the total sample size:

$$\text{expected frequency} = \frac{(\text{row total})(\text{column total})}{\text{sample size}}$$

Because some differences are positive and some are negative, each difference is squared; then each squared difference is divided by the expected frequency. The results for all cells are then summed to produce a statistic that follows the chi-square distribution.

To use this test, the expected frequency for each cell must be greater than 1.0, except for the special case of a two-way table that has two rows and two columns, in which the expected frequency of each cell should be at least 5. (If, for this special case, the expected frequency is less than 5, you can use alternative tests such as Fisher's exact test [see references 2 and 3].)

WORKED-OUT PROBLEM 1 Businesses use a method called *A/B testing* to test different web page designs to see if one design is more effective than another. For one company, designers were interested in the effect of modifying the call-to-action button on the home page. Every visitor to the company's home page was randomly shown either the original call-to-action button (the control) or the new variation. Designers measured success by the download rate: the number of people who downloaded the file divided by the number of people who saw that particular call-to-action button. Results of the experiment yielded the following:

		Call-to-Action Button		
		Original	**New**	**Total**
Download	**Yes**	351	451	802
	No	3,291	3,105	6,396
	Total	3,642	3,556	7,198

The row 1 total shows that 802 visitors to the site downloaded the file. The column 1 total shows 3,642 visitors to the site saw the original call-to-action button. The expected frequency for downloading the new call-to-action button is 396.21, the product of the total number of downloads (802) multiplied by the number of visitors who saw the new call-to-action button (3,556), divided by the total sample size (7,198).

$$\text{expected frequency} = \frac{(802)(3{,}556)}{7{,}198} = 396.21$$

The expected frequencies of the four cells are as follows:

		Call-to-Action Button		
		Original	New	Total
Download	Yes	405.79	396.21	802
	No	3,236.21	3,159.79	6,396
	Total	3,642	3,556	7,198

Spreadsheet results for this study are as follows:

	A		B	C	D
1	**Chi-Square Test**				
2					
3	**Observed Frequencies**				
4			**Call to Action Button**		
5	**Download?**		**Original**	**New**	Total
6		Yes	**351**	**451**	802
7		No	**3291**	**3105**	6396
8		Total	3642	3556	7198
9					
10	Expected Frequencies				
11			Call to Action Button		
12	Download?		Original	New	Total
13		Yes	405.7911	396.2089	802
14		No	3236.2089	3159.7911	6396
15		Total	3642	3556	7198
16					
17	**Data**				
18	**Level of Significance**		**0.05**		
19	Number of Rows		2		
20	Number of Columns		2		
21	Degrees of Freedom		1		
22					
23	**Results**				
24	**Critical Value**		**3.8415**		
25	**Chi-Square Test Statistic**		**16.8527**		
26	***p*-Value**		**0.0000**		
27	**Reject the null hypothesis**				

The results show that the *p*-value for this chi-square test is 0.0000. Because the *p*-value is less than the level of significance α, 0.05, you reject the null hypothesis. You conclude that a relationship exists between the call-to-action button and whether the file is downloaded. You assert that a significant difference exists between the original and new call-to-action buttons in the proportion of visitors who download the file.

WORKED-OUT PROBLEM 2 You decide to use the critical value approach for the study concerning the website design. The computed chi-square statistic is 16.8527 (see the results on the previous page). The number of degrees of freedom for the chi-square test equals the number of rows minus 1 multiplied by the number of columns minus 1:

$$\text{Degrees of freedom} = (\text{Number of rows} - 1) \times (\text{Number of columns} - 1)$$

Using the table of the chi-square distribution (Table C.3), with $\alpha = 0.05$ and the degrees of freedom = $(2 - 1)(2 - 1) = 1$, the critical value of chi-square is equal to 3.841. Because $16.8527 > 3.841$, you reject the null hypothesis.

WORKED-OUT PROBLEM 3 Fast-food chains are evaluated on many variables and the results are summarized periodically in *QSR Magazine*. One important variable is the accuracy of the order. You seek to determine, with a level of significance of $\alpha = 0.05$, whether a difference exists in the proportions of food orders filled correctly at the Burger King, Wendy's, and McDonald's drive-through windows. You use the following data that report the results of drive-through performance.

		Fast Food Chain		
		Burger King	**Wendy's**	**McDonald's**
Order Filled Correctly	**Yes**	203	245	247
	No	43	37	33
	Total	246	282	280

The null and alternative hypotheses are as follows:

H_0: $\pi_1 = \pi_2 = \pi_3$ (No difference exists in the proportion of correct orders among Burger King, Wendy's, and McDonald's.)

H_1: $\pi_1 \neq \pi_2 \neq \pi_3$ (A difference exists in the proportion of correct orders among Burger King, Wendy's, and McDonald's.)

Spreadsheet results for this study are as follows:

	A	B	C	D	E
1	**Chi-Square Test**				
2					
3	**Observed Frequencies**				
4		**Fast Food Chain**			
5	**Order Filled Correctly**	**Burger King**	**Wendy's**	**McDonald's**	Total
6	**Yes**	**203**	**245**	**247**	695
7	**No**	**43**	**37**	**33**	113
8	Total	246	282	280	808
9					
10	Expected Frequencies				
11		Fast Food Chain			
12	Order Filled Correctly	Burger King	Wendy's	McDonald's	Total
13	Yes	211.5965	242.5619	240.8416	695
14	No	34.4035	39.4381	39.1584	113
15	Total	246	282	280	808
16					
17	**Data**				
18	**Level of Significance**	**0.05**			
19	Number of Rows	2			
20	Number of Columns	3			
21	Degrees of Freedom	2			
22					
23	**Results**				
24	**Critical Value**	**5.9915**			
25	**Chi-Square Test Statistic**	**3.7985**			
26	***p*-Value**	**0.1497**			
27	**Do not reject the null hypothesis**				

Because the *p*-value for this chi-square test, 0.1497, is greater than the level of significance α of 0.05, you cannot reject the null hypothesis. Insufficient evidence exists of a difference in the proportion of correct orders filled among Burger King, Wendy's, and McDonald's.

WORKED-OUT PROBLEM 4 Using the critical value approach for the same problem, the computed chi-square statistic is 3.7985 (see the results). At the 0.05 level of significance with the 2 degrees of freedom $[(2-1)(3-1) = 2]$, the chi-square critical value from Table C.3 is 5.991. Because the computed test statistic is less than 5.991, you cannot reject the null hypothesis.

WORKED-OUT PROBLEM 5 The owner of a restaurant serving continental-style entrées wants to determine (at the 0.05 level of significance) whether there is a relationship between the type of dessert and type of entrée ordered during the Friday-to-Sunday time period. Data was collected from 630 customers and organized in the following table.

		Type of Entrée				
		Beef	**Poultry**	**Fish**	**Pasta**	**Total**
Type of Dessert	**Ice Cream**	13	8	12	14	47
	Cake	98	12	29	6	145
	Fruit	8	10	6	2	26
	None	124	98	149	41	412
	Total	243	128	196	63	630

The null and alternative hypotheses are

H_0: No relationship exists between the type of dessert ordered and the type of entree ordered.

H_1: A relationship exists between the type of dessert ordered and the type of entree ordered.

Spreadsheet results are as follows:

	A	B	C	D	E	F
1	**Chi-Square Test**					
2						
3		**Observed Frequencies**				
4		**Type of Entrée**				
5	**Type of Dessert**	**Beef**	**Poultry**	**Fish**	**Pasta**	Total
6	**Ice cream**	**13**	**8**	**12**	**14**	47
7	**Cake**	**98**	**12**	**29**	**6**	145
8	**Fruit**	**8**	**10**	**6**	**2**	26
9	**None**	**124**	**98**	**149**	**41**	412
10	Total	243	128	196	63	630
11						
12		Expected Frequencies				
13		Type of Entrée				
14	Type of Dessert	Beef	Poultry	Fish	Pasta	Total
15	Ice cream	18.1286	9.5492	14.6222	4.7000	47
16	Cake	55.9286	29.4603	45.1111	14.5000	145
17	Fruit	10.0286	5.2825	8.0889	2.6000	26
18	None	158.9143	83.7079	128.1778	41.2000	412
19	Total	243	128	196	63	630
20						
21	**Data**					
22	**Level of Significance**	**0.05**				
23	Number of Rows	4				
24	Number of Columns	4				
25	Degrees of Freedom	9				
26						
27	**Results**					
28	**Critical Value**	**16.9190**				
29	**Chi-Square Test Statistic**	**92.1028**				
30	***p*-Value**	**0.0000**				
31	**Reject the null hypothesis**					

Because the *p*-value for this chi-square test, 0.0000, is less than the level of significance α of 0.05, you reject the null hypothesis. Evidence exists of a relationship between the type of dessert ordered and the type of entree ordered.

WORKED-OUT PROBLEM 6 Using the critical value approach for the same problem, the computed chi-square statistic is 92.1028 (see the previous results). At the 0.05 level of significance with the 9 degrees of freedom [(4 – 1) (4 – 1) = 9], the chi-square critical value from Table C.3 is 16.919. Because the computed test statistic of 92.1028 is greater than 16.919, you reject the null hypothesis.

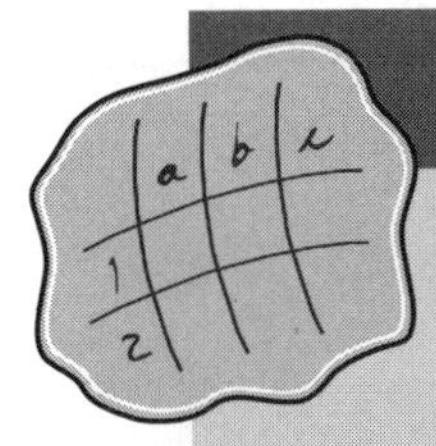

spreadsheet solution

Chi-Square Tests

Chapter 9 Chi-Square contains the spreadsheet that conducts the chi-square test, shown on page 177, for WORKED-OUT PROBLEM 1. Experiment with this spreadsheet by changing the contents of the observed frequencies table in rows 4 through 7, in columns A, B, and C, and by changing the level of significance in cell B18.

Best Practices

Use the **CHISQ.INV.RT(*level of significance, degrees of freedom*)** function to calculate a critical value of the chi-square distribution.

Use the **CHISQ.DIST.RT(*critical value, degrees of freedom*)** function to calculate a probability associated with the chi-square distribution.

How-Tos

Use the spreadsheets in **Chapter 9 Chi-Square Spreadsheets** to perform other chi-square tests that have differing numbers of rows and column categories.

Unlike other Excel files mentioned earlier in this book, this Excel file contains several spreadsheets: ChiSquare2×2 (the one also found in **Chapter 9 Chi-Square**), ChiSquare2×3, ChiSquare3×4, and ChiSquare4×4. To use a particular spreadsheet, click on the sheet tab (near the bottom of the Excel window) for that spreadsheet.

interested in math?

You need the subscripted symbols for the **observed cell frequencies**, f_0, and the **expected cell frequencies**, f_e, to write the equation for the chi-square test for a two-way cross classification table:

$$\chi^2 = \sum_{\text{all cells}} \frac{(f_0 - f_e)^2}{f_e}$$

For the study concerning the manufacture of silicon chips (WORKED-OUT PROBLEM 1), the calculations are as follows:

f_o	f_t	$(f_o - f_t)$	$(f_o - f_t)^2$	$(f_o - f_t)^2/f_t$
351	405.7911	−54.7911	3002.0595	7.3980
3291	3236.209	54.7911	3002.0595	0.9276
451	396.2089	54.7911	3002.0595	7.5770
3105	3159.7911	−54.7911	3002.0595	0.9501
				16.8527

Using the level of significance $\alpha = 0.05$, with $(2 - 1)(2 - 1) = 1$ degree of freedom, from Table C.3, the critical value is 3.841. Because $16.85273 > 3.841$, you reject the null hypothesis.

9.2 One-Way Analysis of Variance (ANOVA): Testing for the Differences Among the Means of More Than Two Groups

Many analyses involve experiments in which you want to test whether differences exist in the means of more than two groups. Evaluating differences between groups is often viewed as a one-factor experiment (also known as a **completely randomized design**) in which the variable that defines the groups is called the factor of interest. A factor of interest can have several *numerical levels* such as baking temperature (e.g., 300°, 350°, 400°, 450°) for an industrial process study, or a factor can have several *categorical levels* such as type of learning materials (Type A, Type B, Type C) for an educational research study.

One-Way ANOVA

CONCEPT The hypothesis test that simultaneously compares the differences among the population means of more than two groups in a one-factor experiment.

INTERPRETATION Unlike the *t* test, which compares differences in two means, the analysis of variance simultaneously compares the differences among the means of more than two groups. Although ANOVA is an acronym for **AN**alysis **O**f **VA**riance, the term is misleading, because the objective in the Analysis of Variance is to analyze differences among the group means, *not* the variances. The null and alternative hypotheses are

H_0: (All the population means are equal.)

H_1: (Not all the population means are equal.)

In ANOVA, the total variation in the values is subdivided into variation that is due to differences among the groups and variation that is due to variation within the groups (see the following figure). Within group variation is called **experimental error**, and the variation between the groups that represents variation due to the factor of interest is called the **treatment effect**.

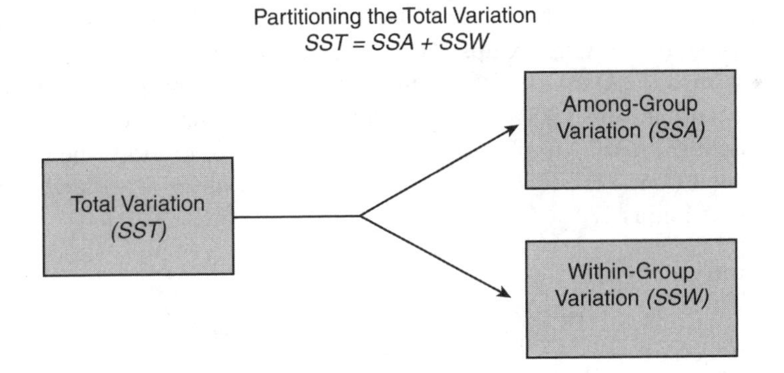

The **sum of squares total (*SST*)** is the total variation that represents the sum of the squared differences between each individual value and the mean of all the values:

$$SST = \text{Sum of (Each value} - \text{Mean of all values)}^2$$

The **sum of squares among groups (*SSA*)** is the among-group variation that represents the sum of the squared differences between the sample mean of each group and the mean of all the values, weighted by the sample size in each group:

$$SSA = \text{Sum of [(Sample size in each group) (Group mean} - \text{Mean of all values)}^2]$$

The **sum of squares within groups** (SSW) is the within-group variation that measures the difference between each value and the mean of its own group and sums the squares of these differences over all groups:

$$SSW = \text{Sum of } [(\text{Each value in the group} - \text{Group mean})^2]$$

The Three Variances of ANOVA

ANOVA derives its name from the fact that the differences between the means of the groups are analyzed by comparing variances. In Section 3.3 on page 45, the variance was calculated as a sum of squared differences around the mean divided by the sample size minus 1.

$$\text{Variance} = S^2 = \frac{\text{Sum of squared differences around the mean}}{\text{sample size - 1}}$$

This sample size minus 1 represents the actual number of values that are free to vary once the mean is known and is called the **degrees of freedom**.

In the analysis of variance, there are three different variances: the variance among groups, the variance within groups, and the total variance. These variances are referred to in the analysis-of-variance terminology as **mean squares**. The **mean square among groups** (MSA) is equal to the sum of squares among groups (SSA) divided by the number of groups minus 1. The **mean square within groups** (MSW) is equal to the sum of squares within groups (SSW) divided by the sample size minus the number of groups. The **mean square total** (MST) is equal to the sum of squares total (SST) divided by the sample size minus 1.

To test the null hypothesis

H_0: All the population means are equal.

against the alternative

H_1: Not all the population means are equal.

you calculate the test statistic F, which follows the F distribution (see Table C.4), as the ratio of two of the variances, MSA to MSW.

$$F = \frac{MSA}{MSW}$$

ANOVA Summary Table

The results of an analysis of variance are usually displayed in an ANOVA summary table. The entries in this table include the sources of variation (among-group, within-group, and total), the degrees of freedom, the sums of squares, the mean squares (or variances), and the F test statistic. A p-value is often included in the table when software is used.

Analysis of Variance Summary Table

Source	Degrees of Freedom	Sum of Squares	Mean Square (Variance)	F
Among groups	number of groups − 1	SSA	$MSA = \frac{SSA}{\text{number of groups} - 1}$	$F = \frac{MSA}{MSW}$
Within groups	sample size − number of groups	SSW	$MSW = \frac{SSW}{\text{sample size} - \text{number of groups}}$	
Total	sample size − 1	SST		

After performing a one-way ANOVA and finding a significant difference among groups, you do not know which groups are significantly different. All that is known is that sufficient evidence exists to state that the population means are not all the same. To determine exactly which groups differ, all possible pairs of groups need to be compared. Many statistical procedures for making these comparisons have been developed (see references 1, 5, 6, 7).

WORKED-OUT PROBLEM 7 You want to determine, with a level of significance $\alpha = 0.05$, whether differences exist among three sets of mathematics learning materials (labeled A, B, and C). You devise an experiment that randomly assigns 24 students to one of the three sets of materials. At the end of a school year, all 24 students are given the same standardized mathematics test that is scored on a 0 to 100 scale. The results of that test are as follows:

Math

Set A	Set B	Set C
87	58	81
80	63	62
74	64	70
82	75	64
74	70	70
81	73	72
97	80	92
71	62	63

Spreadsheet results for these data are as follows:

	A	B	C	D	E	F	G
1	One-Way ANOVA for Learning Materials Study						
2							
3	SUMMARY						
4	*Groups*	*Count*	*Sum*	*Average*	*Variance*		
5	Set A	8	646	80.75	70.21429		
6	Set B	8	545	68.125	56.98214		
7	Set C	8	574	71.75	104.7857		
8							
9							
10	ANOVA						
11	*Source of Variation*	*SS*	*df*	*MS*	*F*	*P-value*	*F crit*
12	Between Groups	676.0833	2	338.0417	4.3716	0.0259	3.4668
13	Within Groups	1623.8750	21	77.3274			
14							
15	Total	2299.9583	23				
16					*Level of significance*		0.05

Because the *p*-value for this test, 0.0259, is less than the level of significance $\alpha = 0.05$, you reject the null hypothesis. You can conclude that the mean scores are not the same for all the sets of mathematics materials. From the results, you see that the mean for materials *A* is 80.75, for materials *B* it is 68.125, and for materials *C* it is 71.75. It appears that the mean score is higher for materials *A* than for materials *B* and *C*.

WORKED-OUT PROBLEM 8 Using the critical value approach for the same problem, the computed *F* statistic is 4.37 (see the earlier results). To determine the critical value of *F*, you refer to the table of the *F* statistic (Table C.4). This table requires these degrees of freedom:

- The numerator degrees of freedom, equal to the number of groups minus 1
- The denominator degrees of freedom, equal to the sample size minus the number of groups

With three groups and a sample size of 24, the numerator degrees of freedom are 2 ($3 - 1 = 2$), and the denominator degrees of freedom are 21 ($24 - 3 = 21$). With the level of significance $\alpha = 0.05$, the critical value of *F* from Table C.4 is 3.47 (also shown in the spreadsheet results). Because the decision rule is to reject H_0 if $F >$ critical value of *F*, and $F = 4.37 > 3.47$, you reject the null hypothesis.

interested in math?

To form the equations for the three mean squares and the test statistic F, you assemble these symbols:

- $\bar{\bar{X}}$, pronounced as "X double bar," that represents the overall or grand mean
- a subscripted X Bar, $\bar{X}_j$, that represents the mean of a group
- a double-subscripted uppercase italic X, X_{ij}, that represents individual values in group j
- a subscripted lowercase italic n, n_j, that represents the sample size in a group
- a lowercase italic n, n, that represents the total sample size (sum of the sample sizes of each group)
- a lowercase italic c, c, that represents the number of groups

First, you form the equation for the grand mean as

$$\bar{\bar{X}} = \frac{\sum_{j=1}^{c}\sum_{i=1}^{n_j} X_{ij}}{n} = \text{grand mean}$$

$\bar{X}_j$ = sample mean of group j

X_{ij} = ith value in group j

n_j = number of values in group j

n = total number of values in all groups combined
(that is, $n = n_1 + n_2 + \cdots + n_c$)

c = number of groups of the factor of interest

With $\bar{\bar{X}}$ defined, you then form the equations that define the sum of squares total, SST, the sum of squares among groups, SSA, and the sum of squares within groups, SSW:

$$SST = \sum_{j=1}^{c}\sum_{i=1}^{n_j}(X_{ij} - \bar{\bar{X}})^2$$

$$SSA = \sum_{j=1}^{c} n_j(\bar{X}_j - \bar{\bar{X}})^2$$

$$SSW = \sum_{j=1}^{c}\sum_{i=1}^{n_j}(X_{ij} - \bar{X}_j)^2$$

(continues)

Next, using these definitions, you form the equations for the mean squares:

$$MSA = \frac{SSA}{\text{number of groups - 1}}$$

$$MSW = \frac{SSW}{\text{sample size - number of groups}}$$

$$MST = \frac{SST}{\text{sample size - 1}}$$

Finally, using the definitions of *MSA* and *MSW*, you form the equation for the test statistic *F*:

$$F = \frac{MSA}{MSW}$$

Using the data from WORKED-OUT PROBLEM 7,

$$\bar{\bar{X}} = \frac{1,765}{24} = 73.5417$$

$$SST = \sum_{j=1}^{c}\sum_{i=1}^{n_j}(X_{ij} - \bar{\bar{X}})^2 = (87 - 73.5417)^2 + \cdots + (71 - 73.5417)^2$$
$$+ (58 - 73.5417)^2 + \cdots + (62 - 73.5417)^2$$
$$+ (81 - 73.5417)^2 + \cdots + (63 - 73.5417)^2 = 2,299.9583$$

$$SSA = \sum_{j=1}^{c} n_j(\bar{X}_j - \bar{\bar{X}})^2 = 8(80.75 - 73.5417)^2$$
$$+ 8(68.125 - 73.5417)^2 + 8(71.75 - 73.5417)^2$$
$$= 676.0833$$

$$SSW = \sum_{j=1}^{c}\sum_{i=1}^{n_j}(X_{ij} - \bar{X}_j)^2 = (87 - 80.75)^2 + \cdots + (71 - 80.75)^2$$
$$+ (58 - 68.125)^2 + \cdots + (62 - 68.125)^2$$
$$+ (81 - 71.75)^2 + \cdots + (63 - 71.75)^2 = 1,623.875$$

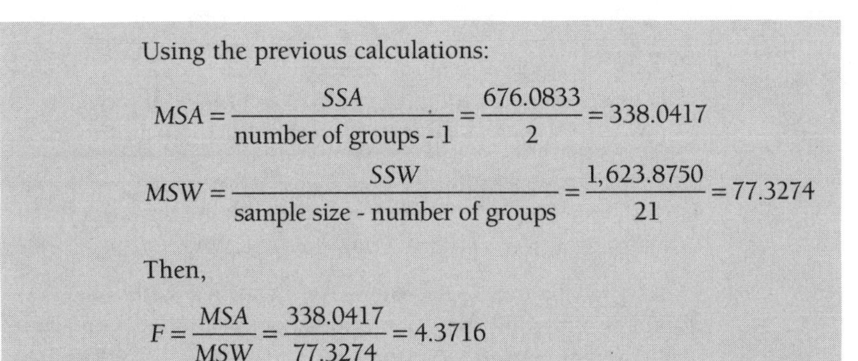

Using the previous calculations:

$$MSA = \frac{SSA}{\text{number of groups - 1}} = \frac{676.0833}{2} = 338.0417$$

$$MSW = \frac{SSW}{\text{sample size - number of groups}} = \frac{1{,}623.8750}{21} = 77.3274$$

Then,

$$F = \frac{MSA}{MSW} = \frac{338.0417}{77.3274} = 4.3716$$

Because $F = 4.3716 > 3.47$, you reject H_0.

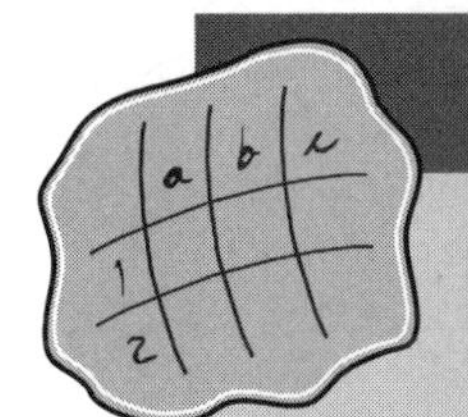

spreadsheet solution

One-Way ANOVA

Chapter 9 One-Way ANOVA contains the WORKED-OUT PROBLEM 7 one-way ANOVA for the learning materials study shown on page 186. Experiment with this spreadsheet by changing the level of significance in cell G16.

Best Practices

Use the **COUNT**, **SUM**, **AVERAGE**, and **VAR.S** functions to calculate the counts, sums, averages, and variances, respectively.

Use the **F.INV.RT(level of significance, degrees of freedom between groups, degrees of freedom within groups)** function to calculate the critical value of the F distribution.

Use the **F.DIST.RT(F test statistic, degrees of freedom between groups, degrees of freedom within groups)** function to calculate the p-value.

Use the **DEVSQ(*cell range of the one-way ANOVA data*)** function to compute the sum of the squares for the total variation (*SST*).

(*continues*)

How Tos

To help calculate the *SSA*, use the **DEVSQ** function for each group's cell range and subtract the values calculated from the *SST*.

Tip ATT5 in Appendix E describes using the Analysis ToolPak as a second way to perform a one-way ANOVA.

Modifying **Chapter 9 One-Way ANOVA** for use with other problems is beyond the scope of this book. However, you can compare that spreadsheet to the one in **Chapter 9 WORKED-OUT PROBLEM 9** to discover the modifications to formulas necessary in both the spreadsheet SUMMARY and ANOVA tables.

WORKED-OUT PROBLEM 9 A pet food company is looking to expand its product line beyond its current kidney and shrimp-based cat foods. The company developed two new products, one based on chicken livers, and the other based on salmon. The company conducted an experiment to compare the two new products with its two existing ones as well as a generic beef-based product sold in a supermarket chain.

For the experiment, a sample of 50 cats from the population at a local animal shelter was selected. Ten cats were randomly assigned to each of the five products being tested. Each of the cats was then presented with three ounces of the selected food in a dish at feeding time. The researchers defined the variable to be measured as the number of ounces of food that the cat consumed within a 10-minute time interval that began when the filled dish was presented. The results for this experiment are summarized in the following table.

CatFood

Kidney	Shrimp	Chicken Liver	Salmon	Beef
2.37	2.26	2.29	1.79	2.09
2.62	2.69	2.23	2.33	1.87
2.31	2.25	2.41	1.96	1.67
2.47	2.45	2.68	2.05	1.64
2.59	2.34	2.25	2.26	2.16
2.62	2.37	2.17	2.24	1.75
2.34	2.22	2.37	1.96	1.18
2.47	2.56	2.26	1.58	1.92
2.45	2.36	2.45	2.18	1.32
2.32	2.59	2.57	1.93	1.94

Spreadsheet results for this study are as follows:

	A	B	C	D	E	F	G
1	**One-Way ANOVA for Cat Food Study**						
2							
3	**SUMMARY**						
4	*Groups*	*Count*	*Sum*	*Average*	*Variance*		
5	Kidney	10	24.56	2.456	0.0148		
6	Shrimp	10	24.09	2.409	0.0253		
7	Chicken Liver	10	23.68	2.368	0.0263		
8	Salmon	10	20.28	2.028	0.0544		
9	Beef	10	17.54	1.754	0.0990		
10							
11							
12	**ANOVA**						
13	*Source of Variation*	*SS*	*df*	*MS*	*F*	*P-value*	*F crit*
14	Between Groups	3.6590	4	0.9147	20.8054	0.0000	2.5787
15	Within Groups	1.9785	45	0.0440			
16							
17	Total	5.6375	49				
18					*Level of significance*		0.05

Because the *p*-value for this test, 0.000, is less than the level of significance $\alpha = 0.05$, you reject the null hypothesis. You conclude that evidence of a difference exists in the mean amount of food eaten among the five types of cat foods.

WORKED-OUT PROBLEM 10 Using the critical value approach for the same problem, the computed *F* statistic is 20.8054. At the level of significance $\alpha = 0.05$, with 4 degrees of freedom in the numerator (5 – 1) and 45 degrees of freedom in the denominator (50 – 5), the critical value of *F* from the Excel results is 2.5787. Because the computed *F* test statistic 20.8054 is greater than 2.5787, you reject the null hypothesis.

One-Way ANOVA Assumptions

There are three major assumptions you must make to use the one-way ANOVA *F* test: randomness and independence, normality, and homogeneity of variance.

The first assumption, randomness and independence, always must be met, because the validity of your experiment depends on the random sampling or random assignment of items or subjects to groups. Departures from this assumption can seriously affect inferences from the analysis of variance. These problems are discussed more thoroughly in references 6 and 7.

The second assumption, normality, states that the values in each group are selected from normally distributed populations. The one-way ANOVA *F* test is not very sensitive to departures from this assumption of normality. As long as the distributions are not very skewed, the level of significance of the ANOVA *F* test is usually not greatly affected by lack of normality, particularly for large samples. When only the normality assumption is seriously violated, nonparametric alternatives to the one-way ANOVA *F* test are available (see references 1 and 5).

The third assumption, equality of variances, states that the variance within each population should be equal for all populations. Although the one-way ANOVA *F* test is relatively robust or insensitive with respect to the assumption of equal group variances, large departures from this assumption can seriously affect the level of significance and the power of the test. Therefore, various procedures have been developed to test the assumption of homogeneity of variance (see references 1, 4, and 5).

One way to evaluate the assumptions is to construct a side-by-side box-and-whisker plot of the groups to study their central tendency, variation, and shape.

Other Experimental Designs

The one-way analysis of variance is the simplest type of experimental design, because it considers only one factor of interest. More complicated experimental designs examine at least two factors of interest simultaneously. For more information on these designs, see references 1, 4, 6, and 7.

Important Equations

Chi-square test for a two-way cross-classification table:

(9.1) $$\chi^2 = \sum_{all\ cells} \frac{(f_0 - f_e)^2}{f_e}$$

ANOVA calculations:

(9.2) $$SST = \sum_{j=1}^{c} \sum_{i=1}^{n_j} (X_{ij} - \bar{\bar{X}})^2$$

(9.3) $$SSA = \sum_{j=1}^{c} n_j (\bar{X}_j - \bar{\bar{X}})^2$$

(9.4) $$SSW = \sum_{j=1}^{c} \sum_{i=1}^{n_j} (X_{ij} - \bar{X}_j)^2$$

(9.5) $$MSA = \frac{SSA}{\text{number of groups - 1}}$$

(9.6) $$MSW = \frac{SSW}{\text{sample size - number of groups}}$$

(9.7) $$MST = \frac{SST}{\text{sample size - 1}}$$

(9.8) $$F = \frac{MSA}{MSW}$$

One-Minute Summary

Tests for the differences among more than two groups:

- If your data are categorical, use chi-square (χ^2) tests (can also use for two groups).
- If your data are numerical and if you have one factor, use the one-way ANOVA.

Test Yourself

Short Answers

1. In a one-way ANOVA, if the F test statistic is greater than the critical F value, you:
   (a) reject H_0 because there is evidence all the means differ
   (b) reject H_0 because there is evidence at least one of the means differs from the others
   (c) do not reject H_0 because there is no evidence of a difference in the means
   (d) do not reject H_0 because one mean is different from the others
2. In a one-way ANOVA, if the p-value is greater than the level of significance, you:
   (a) reject H_0 because there is evidence all the means differ
   (b) reject H_0 because there is evidence at least one of the means differs from the others.
   (c) do not reject H_0 because there is insufficient evidence of a difference in the means
   (d) do not reject H_0 because one mean is different from the others
3. The F test statistic in a one-way ANOVA is:
   (a) MSW/MSA
   (b) SSW/SSA
   (c) MSA/MSW
   (d) SSA/SSW

4. In a one-way ANOVA, the null hypothesis is always:
   (a) all the population means are different
   (b) some of the population means are different
   (c) some of the population means are the same
   (d) all of the population means are the same

5. A car rental company wants to select a computer software package for its reservation system. Three software packages (*A*, *B*, and *C*) are commercially available. The car rental company will choose the package that has the lowest mean number of renters for whom a car is not available at the time of pickup. An experiment is set up in which each package is used to make reservations for five randomly selected weeks. How should the data be analyzed?
   (a) Chi-square test for differences in proportions
   (b) One-way ANOVA *F* test
   (c) *t* test for the differences in means
   (d) *t* test for the mean difference

The following should be used to answer Questions 6 through 9:

For fast-food restaurants, the drive-through window is an increasing source of revenue. The chain that offers that fastest service is considered most likely to attract additional customers. In a study of 20 drive-through times (from menu board to departure) at 5 fast-food chains, the following ANOVA table was developed.

Source	DF	Sum of Squares	Mean Squares	*F*
Among Groups (Chains)		6,536	1,634.0	12.51
Within Groups (Chains)	95		130.6	
Total	99	18,943		

6. Referring to the preceding table, the Among Groups degrees of freedom is:
   (a) 3
   (b) 4
   (c) 12
   (d) 16

7. Referring to the preceding table, the within groups sum of squares is:
   (a) 12,407
   (b) 95
   (c) 130.6
   (d) 4

8. Referring to the preceding table, the within groups mean squares is:
   (a) 12,407
   (b) 95
   (c) 130.6
   (d) 4

9. Referring to the preceding table, at the 0.05 level of significance, you:
   (a) do not reject the null hypothesis and conclude that no difference exists in the mean drive-up time between the fast-food chains
   (b) do not reject the null hypothesis and conclude that a difference exists in the mean drive-up time between the fast-food chains
   (c) reject the null hypothesis and conclude that a difference exists in the mean drive-up time between the fast-food chains
   (d) reject the null hypothesis and conclude that no difference exists in the mean drive-up time between the fast-food chains

10. When testing for independence in a contingency table with three rows and four columns, there are ________ degrees of freedom.
    (a) 5
    (b) 6
    (c) 7
    (d) 12

11. In testing a hypothesis using the chi-square test, the theoretical frequencies are based on the:
    (a) null hypothesis
    (b) alternative hypothesis
    (c) normal distribution
    (d) t distribution

12. An agronomist is studying three different varieties of tomato to determine whether a difference exists in the proportion of seeds that germinate. Random samples of 100 seeds of each of three varieties are subjected to the same starting conditions. How should the data be analyzed?
    (a) Chi-square test for differences in proportions
    (b) One-way ANOVA F test
    (c) t test for the differences in means
    (d) t test for the mean difference

Answer True or False:

13. A test for the difference between two proportions can be performed using the chi-square distribution.

14. The one-way analysis-of-variance (ANOVA) tests hypotheses about the difference between population proportions.
15. The one-way analysis-of-variance (ANOVA) tests hypotheses about the difference between population means.
16. The one-way analysis-of-variance (ANOVA) tests hypotheses about the difference between population variances.
17. The Mean Squares in an ANOVA can never be negative.
18. In a one-factor ANOVA, the Among sum of squares and Within sum of squares must add up to the total sum of squares.
19. If you use the chi-square method of analysis to test for the difference between two proportions, you must assume that there are at least five observed frequencies in each cell of the contingency table.
20. If you use the chi-square method of analysis to test for independence in a contingency table with more than two rows and more than two columns, you must assume that there is at least one theoretical frequency in each cell of the contingency table.

Answers to Test Yourself Short Answers

1. b
2. c
3. c
4. d
5. b
6. b
7. a
8. c
9. c
10. b
11. a
12. a
13. True
14. False
15. True
16. False
17. True
18. True
19. True
20. True

Problems

1. Are men and women equally likely to say that a major reason they use Facebook is to share with many people at once? A survey reported that 42% of men (193 out of 459 sampled) and 50% of women (250 out of 501 sampled) said that a major reason they use Facebook is to share with many people at once. (Data extracted from "6 new facts about Facebook," **bit.ly/1rCTrOO**.)

(a) At the 0.05 level of significance, is there evidence of a difference between men and women in the proportion that say that a major reason they use Facebook is to share with many people at once?

(b) Compare the results of (a) with those of Problem 1 of Chapter 8, on page 170.

2. Do people trust banks to do what is right? A survey done in the United States and Japan revealed that, of 500 respondents in the United States, 250 said that they trusted banks to do what is right; in Japan, of 200 respondents, 120 said that they trusted banks to do what is right. (Data extracted from F. Norris, "Where Banking Crisis Raged, Trust Is Slow to Return," *The New York Times*, 26 January 2013, p. B3.)

   (a) At the 0.05 level of significance, is there evidence of a difference between respondents in the United States and Japan who trust banks to do what is right?

   (b) Compare the results of (a) with those of Problem 3 of Chapter 8, on page 171.

3. More shoppers do the majority of their grocery shopping on Saturday than any other day of the week. However, is there a difference in the various age groups in the proportion of people who do the majority of their grocery shopping on Saturday?

   A study reported that, of 200 shoppers under 35 years old, 56 did their major shopping on Saturday; of 200 shoppers between 35 and 54 years old, 80 did their major shopping on Saturday; and of 200 shoppers over 54 years old, 32 did their major shopping on Saturday, At the 0.05 level of significance, is there evidence of a difference among the age groups with respect to major grocery shopping day?

4. You seek to determine, with a level of significance of $\alpha = 0.05$, whether there was a relationship between numbers selected for the Vietnam War era military draft lottery system and the time of the year a man was born. The following shows how many low (1–122), medium (123–244), and high (245–366) numbers were drawn for birth dates in each quarter of the year.

		Quarter of Year				
		Jan–Mar	**Apr–Jun**	**Jul–Sep**	**Oct–Dec**	**Total**
	Low	21	28	35	38	122
Number Set	**Medium**	34	22	29	37	122
	High	36	41	28	17	122
	Total	91	91	92	92	366

At the 0.05 level of significance, is there a relationship between draft number and quarter of the year?

5. You want to determine, with a level of significance $\alpha = 0.05$, whether differences exist among the four plants that fill boxes of a particular brand of cereal. You select samples of 20 cereal boxes from each of the four plants. The weights of these cereal boxes (in grams) are as follows.

BoxFills

Plant 1		Plant 2		Plant 3		Plant 4	
361.43	364.78	370.26	360.27	367.53	390.12	361.95	369.36
368.91	376.75	357.19	362.54	388.36	335.27	381.95	363.11
365.78	353.37	360.64	352.22	359.33	366.37	383.90	400.18
389.70	372.73	398.68	347.28	367.60	371.49	358.07	358.61
390.96	363.91	380.86	350.43	358.06	358.01	382.40	370.87
372.62	375.68	334.95	376.50	369.93	373.18	386.20	380.56
390.69	380.98	359.26	369.27	355.84	377.40	373.47	376.21
364.93	354.61	389.56	377.36	382.08	396.30	381.16	380.97
387.13	378.03	371.38	368.50	381.45	354.82	379.41	365.78
360.77	374.24	373.06	363.86	356.20	383.78	382.01	395.55

What conclusions can you reach?

6. *QSR* reports on the largest quick-serve and fast casual restaurants in the United States. Do the various market segments (burger, chicken, sandwich, and pizza) differ in their mean sales per unit? The following table presents the mean sales per unit by segment in a recent year.

FastFood Chain

Burger	Chicken	Sandwich	Pizza
2600.0	3158.0	481.0	883.0
1483.8	957.0	2427.2	710.2
1195.0	1242.0	993.2	829.0
1470.0	1717.5	878.8	465.0
903.4	1475.0	345.0	574.9
1254.2	1184.0	2556.4	915.0

Source: Data extracted from "The QSR 50," **bit.ly/1mw56xA**.

At the 0.05 level of significance, is there evidence of a difference in the mean sales per unit among the market segments?

7. A sporting goods manufacturing company wanted to compare the distance traveled by golf balls produced using each of four different designs. Ten balls were manufactured with each design and each ball was tested at a testing facility using a robot to hit the balls. The results (distance traveled in yards) for the four designs were as follows:

GolfBall

Design1	Design2	Design3	Design4
206.32	217.08	226.77	230.55
207.94	221.43	224.79	227.95
206.19	218.04	229.75	231.84
204.45	224.13	228.51	224.87
209.65	211.82	221.44	229.49
203.81	213.90	223.85	231.10
206.75	221.28	223.97	221.53
205.68	229.43	234.30	235.45
204.49	213.54	219.50	228.35
210.86	214.51	233.00	225.09

At the 0.05 level of significance, is there evidence of a difference in the mean distances traveled by the golf balls with different designs?

Answers to Test Yourself Problems

1. (a) Because the *p*-value for this chi-square test, 0.0148, is less than the level of significance α of 0.05 (or the chi-square statistic = 5.9432 > 3.841), you reject the null hypothesis. There is evidence of a difference between men and women in the proportion of people that say that a major reason they use Facebook is sharing with many people at once.
   (b) The results are the same because the chi-square statistic with one degree of freedom is the square of the Z statistic.

2. (a) Because the *p*-value for this chi-square test, 0.0166, is less than the level of significance α of 0.05 (or the chi-square statistic = 5.733 > 3.841), you reject the null hypothesis. There is evidence of a difference in the proportion of U.S. and Japanese respondents who trusted banks to do what is right.
   (b) The results are the same because the chi-square statistic with one degree of freedom is the square of the Z statistic.

3. Because the *p*-value for this chi-square test, 0.0000, is less than the level of significance α of 0.05 (or the chi-square statistic = 28.5714 > 5.991),

you reject the null hypothesis. There is evidence of a difference among the age groups with respect to the major grocery shopping day.

4. Because the *p*-value for this chi-square test, 0.0021, is less than the level of significance α of 0.05 (or the chi-square statistic = 20.6804 > 12.5916), you reject the null hypothesis. Evidence exists of a relationship between the number selected and the time of the year in which the man was born. It appears that men who were born between January and June were more likely than expected to have high numbers, whereas men born between July and December were more likely than expected to have low numbers.
5. Because the *p*-value for this test, 0.0959, is greater than the level of significance $\alpha = 0.05$ (or the computed *F* test statistic = 2.1913 is less than the critical value of $F = 2.725$), you cannot reject the null hypothesis. You conclude that there is insufficient evidence of a difference in the mean cereal weights among the four plants.
6. Because the *p*-value for this test, 0.1630, is greater than the level of significance $\alpha = 0.05$ (or the computed *F* test statistic 1.895 is less than the critical value of $F = 3.0984$), you cannot reject the null hypothesis. You conclude that there is insufficient evidence of a difference in the mean sales per unit among the market segments.
7. Because the $p\text{-value} = 0.0000 < 0.05$ (or $F = 53.03 > 2.92$), reject H_0. There is evidence of a difference in the mean distances traveled by the golf balls with different designs.

References

1. Berenson, M. L., D. M. Levine, and K. A. Szabat. *Basic Business Statistics: Concepts and Applications*, Thirteenth Edition. Upper Saddle River, NJ: Pearson Education, 2015.
2. Conover, W. J. *Practical Nonparametric Statistics*, Third Edition. New York: Wiley, 2000.
3. Daniel, W. *Applied Nonparametric Statistics*, Second Edition. Boston: Houghton Mifflin, 1990.
4. Levine, D. M. *Statistics for Six Sigma Green Belts Using Minitab and JMP*. Upper Saddle River, NJ: Prentice Hall, 2006.
5. Levine, D. M., D. Stephan, and K. A. Szabat. *Statistics for Managers Using Microsoft Excel*, Seventh Edition. Upper Saddle River, NJ: Pearson Education, 2014.
6. Montgomery, D. C. *Design and Analysis of Experiments*, Sixth Edition. New York: John Wiley, 2005.
7. Kutner, M. H., C. Nachtsheim, J. Neter, and W. Li. *Applied Linear Statistical Models*, Fifth Edition., New York: McGraw-Hill-Irwin, 2005.
8. Microsoft Excel 2013. Redmond, WA: Microsoft Corporation, 2012.

CHAPTER 10

Simple Linear Regression

Regression methods are inferential methods that allow you to use the values of one variable to predict the values of another variable. For example, managers of a growing chain of retail stores might wonder if larger-sized stores generate greater sales, farmers might want to predict the weight of a pumpkin based on its circumference, and baseball fans might want to predict the number of games a team wins in a season based on the number of runs the team scores.

In this chapter, you learn the basics of regression analysis and the specifics about *simple linear* regression models that examine the relationship between numerical variables.

10.1 Basics of Regression Analysis

Learning regression analysis requires understanding the several new concepts and vocabulary terms that form the basis for regression analysis.

Prediction

CONCEPT The term used to describe what a regression model does.

INTERPRETATION Prediction, in the statistical sense, is the explanation of how the values of one or more variables are related to another variable. That explanation takes the form of a mathematical expression. For example, for a retail store chain, it might be that increasing the size of an individual store by one-third increases sales at that store by 50 percent for the range of store sizes from 3,000 to 7,000 square feet. If a 3,000-square-foot store had sales of $10,000 per week, you could *predict* that increasing the size of the store to 4,000 square feet would probably result in sales of about $15,000 per week.

Statistical prediction is not psychic guessing, nor is it a statement about future events. When choosing independent variables for prediction, there must be a logical relationship between the independent variables and the dependent variable. Many relationships among logically unrelated variables happen by chance. For example, in one famous example, a researcher showed how increasing the imports of Mexican lemons decreased U.S. highway fatalities: for every increase of about 300 metric tons of lemons imported, there was a drop of 1 fatality per 100,000 people in the rate of fatalities (see reference 4). Such a non-logical relationship is called a spurious correlation.

Dependent Variable Y

CONCEPT The variable that a regression model seeks to predict.

INTERPRETATION The dependent variable is also known as the **response variable**.

Independent Variable X

CONCEPT A variable that a regression model seeks to use to predict the dependent variable.

EXAMPLE The independent variable store size being used to predict the dependent variable store sales in the example given in the "Interpretation" section for "Prediction."

INTERPRETATION The independent variable is also known as the **explanatory variable**.

Scatter Plots and Regression Analysis

Scatter plots can visualize the relationship in the special case of one independent variable *X* and one dependent variable *Y*. Scatter plots can suggest a specific type of relationship between the *X* and *Y* variables that regression analysis can later explore. For example, Panels A though F on the next page show six different kinds of patterns.

The patterns seen can be described as follows:

- Panel A, positive straight-line or linear relationship between *X* and *Y*. As the value of the *X* variable increases, the value of the *Y* variable increases.

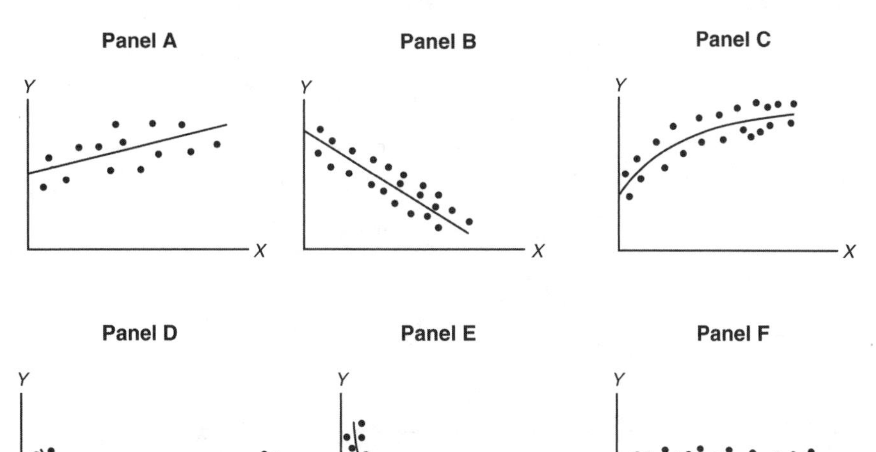

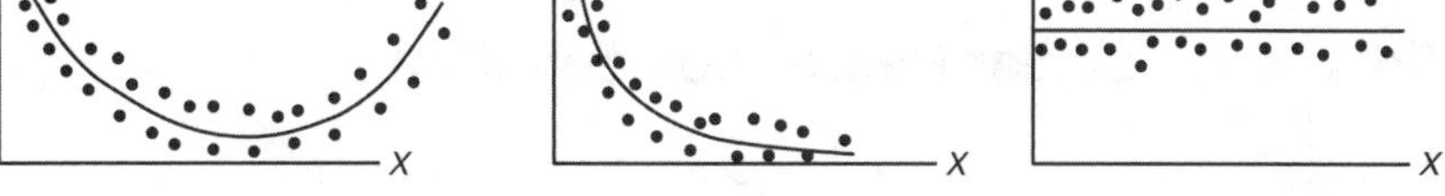

- Panel B, negative straight-line or linear relationship between *X* and *Y*.
- Panel C, a positive curvilinear relationship between *X* and *Y*. The values of *Y* are increasing as *X* increases, but this increase tapers off beyond certain values of *X*.
- Panel D, a U-shaped relationship between *X* and *Y*. As *X* increases, at first *Y* decreases. However, as *X* continues to increase, *Y* not only stops decreasing but actually increases above its minimum value.
- Panel E, an exponential relationship between *X* and *Y*. In this case, *Y* decreases very rapidly as *X* first increases, but then decreases much less rapidly as *X* increases further.
- Panel F, little or no pattern to the relationship between *X* and *Y*. The same or similar values of *Y* are associated with each value of *X*. An *X* value in this data set cannot be used to predict a *Y* value.

Simple Linear Regression

CONCEPT The regression model that uses a straight-line (linear) relationship to predict a *numerical* dependent variable *Y* from a single *numerical* independent variable *X*.

INTERPRETATION Two values define the straight-line relationship: the **Y intercept** and the **slope**.

The *Y* intercept is the value of *Y* when $X = 0$ (see figure on the next page); the slope is the change in *Y* (the vertical dashed line in the figure on the next page) per unit change in *X* (horizontal dashed line). Positive straight-line relationships (Panel A above) have a positive slope; negative straight-line relationships (Panel B above) have a negative slope.

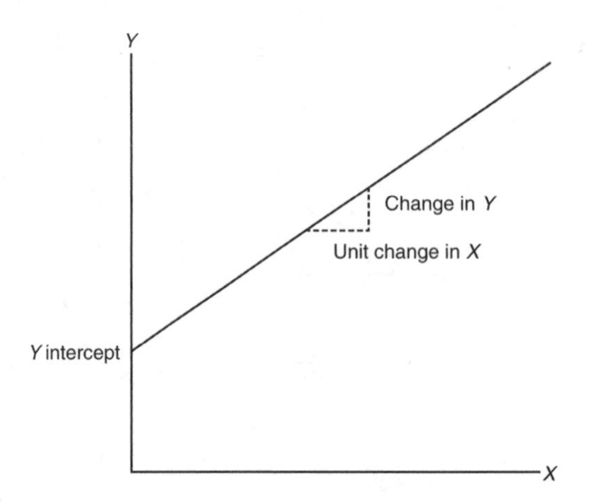

The Simple Linear Regression Equation

CONCEPT The equation that represents the straight-line regression model that can be used for prediction. In this equation, the symbol b_0 represents the Y intercept and the symbol b_1 represents the slope.

INTERPRETATION The general form of the simple linear regression equation is $Y = b_0 + b_1X$. This means that, for a given value of X, that value is multiplied by b_1 (the slope) and the product that results is added to b_0 (the Y intercept) to calculate the predicted Y value. A simple linear regression equation is valid only for the range of X values that was used to develop the regression model (as the next section explains). Note that sometimes the general form is written as $Y = a + bX$, in which the symbol a represents the Y intercept and the symbol b represents the slope.

WORKED-OUT PROBLEM 1 Consider the simple linear regression model defined by the equation $Y = 100 + 1.5X$ for the range of X values from 10 to 50. Predict the Y value that would be associated with the X value 40. The predicted Y value would be 160 ($100 + 1.5 \times 40$).

10.2 Developing a Simple Linear Regression Model

Developing a simple linear regression model requires two tasks: using a calculational method to determine the most appropriate equation and performing a residual analysis that evaluates whether a linear model is the most appropriate type of model to use.

Least-Squares Method

CONCEPT The calculational method that minimizes the sum of the squared differences between the actual values of the dependent variable Y and the predicted values of Y.

INTERPRETATION For plotted sets of X and Y values, there are many possible straight lines, each with its own values of b_0 and b_1, that might seem to fit the data. The least-squares method finds the values for the Y intercept and the slope that makes the sum of the squared differences between the actual values of the dependent variable Y and the predicted values of Y as small as possible.

Calculating the Y intercept and the slope using the least-squares method is tedious and can be subject to rounding errors if you use a simple four-function calculator. You can get more accurate results faster if you use regression software routines to perform the calculations.

WORKED-OUT PROBLEM 2 You want to assist a moving company owner to develop a more accurate method of predicting the labor hours needed to move a volume of goods (measured in cubic feet). The manager has collected the following data for 36 moves and has eliminated the travel-time portion of the time needed for the move.

Moving

Hours	Cu. Feet	Hours	Cu. Feet	Hours	Cu. Feet
24.00	545	19.50	344	37.00	757
13.50	400	18.00	360	32.00	600
26.25	562	28.00	750	34.00	796
25.00	540	27.00	650	25.00	577
9.00	220	21.00	415	31.00	500
20.00	344	15.00	275	24.00	695
22.00	569	25.00	557	40.00	1,054
11.25	340	45.00	1,028	27.00	486
50.00	900	29.00	793	18.00	442
12.00	285	21.00	523	62.50	1,249
38.75	865	22.00	564	53.75	995
40.00	831	16.50	312	79.50	1,397

The scatter plot for these data (shown below) suggests a positive straight-line relationship between cubic feet moved (X) and labor hours (Y). As the cubic footage moved increases, labor hours increase approximately as a straight line.

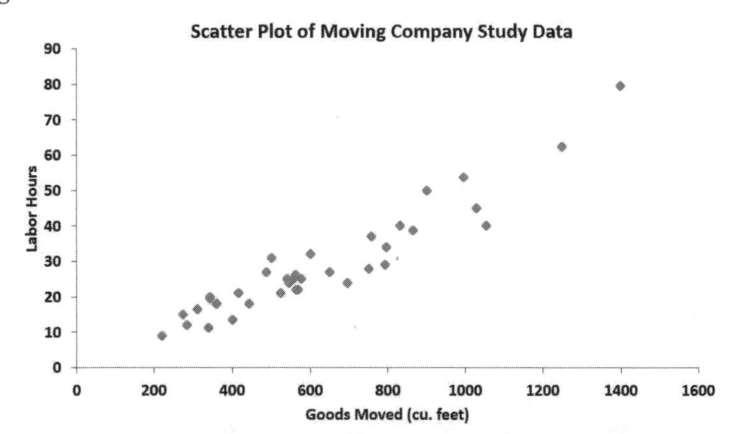

Spreadsheet least-squares method results for this study are as follows:

	A	B	C	D	E	F	G	H	I
1	Simple Linear Regression								
2									
3	*Regression Statistics*								
4	Multiple R	0.9430							
5	R Square	0.8892							
6	Adjusted R Square	0.8860							
7	Standard Error	5.0314							
8	Observations	36							
9									
10	ANOVA								
11		*df*	*SS*	*MS*	*F*	*Significance F*			
12	Regression	1	6910.7189	6910.7189	272.9864	0.0000			
13	Residual	34	860.7186	25.3153					
14	Total	35	7771.4375						
15									
16		*Coefficients*	*Standard Error*	*t Stat*	*P-value*	*Lower 95%*	*Upper 95%*	*Lower 95%*	*Upper 95%*
17	Intercept	-2.3697	2.0733	-1.1430	0.2610	-6.5830	1.8437	-6.5830	1.8437
18	Goods Moved	0.0501	0.0030	16.5223	0.0000	0.0439	0.0562	0.0439	0.0562

The results show that $b_1 = 0.05$ and $b_0 = -2.37$. Thus, the equation for the best straight line for these data is this:

$$\text{Predicted value of labor hours} = -2.37 + 0.05 \times \text{cubic feet moved}$$

The slope b_1 was computed as +0.05. This means that for each increase of 1 unit in X, the value of Y is estimated to increase by 0.05 units. In other words, for each increase of 1 cubic foot to be moved, the fitted model predicts that the labor hours are estimated to increase by 0.05 hours.

The Y intercept b_0 was computed to be –2.37. The Y intercept represents the value of Y when X equals 0. Because the cubic feet moved cannot be 0, the Y intercept has no practical interpretation. The sample linear regression line for these data, plotted with the actual values, is:

Prediction Using a Simple Linear Regression Model

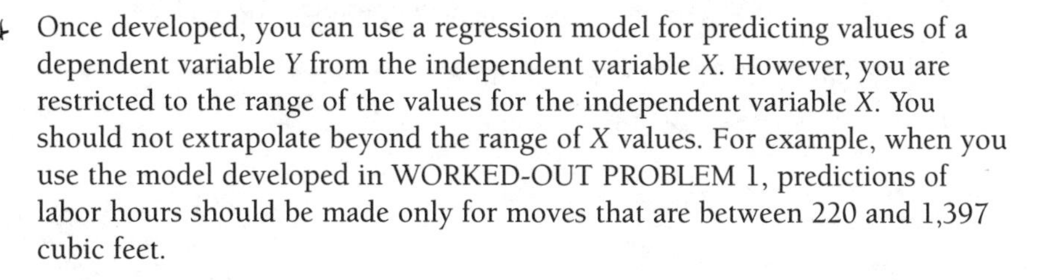

important point

Once developed, you can use a regression model for predicting values of a dependent variable Y from the independent variable X. However, you are restricted to the range of the values for the independent variable X. You should not extrapolate beyond the range of X values. For example, when you use the model developed in WORKED-OUT PROBLEM 1, predictions of labor hours should be made only for moves that are between 220 and 1,397 cubic feet.

WORKED-OUT PROBLEM 3 Using the regression model developed in WORKED-OUT PROBLEM 2, you want to predict the labor hours for a moving job that consists of 800 cubic feet. You predict that the labor hours for a move would be 37.69 ($-2.3697 + 0.0501 \times 800$).

equation blackboard (optional)

interested in math?

You use the symbols for the Y intercept, b_0, and the slope, b_1, the sample size, n, and these symbols:

- the subscripted YHat, $\hat{Y}_i$, for predicted Y values
- the subscripted italic capital X for the independent X values
- the subscripted italic capital Y for the dependent Y values
- $\bar{X}$ for the mean of the X values
- $\bar{Y}$ for the mean of the Y values

to write the equation for a simple linear regression model:

$$\hat{Y}_i = b_0 + b_1 X_i$$

You use that equation and these summations:

- $\sum_{i=1}^{n} X_i$, the sum of the X values
- $\sum_{i=1}^{n} Y_i$, the sum of the Y values
- $\sum_{i=1}^{n} X_i^2$, the sum of the squared X values

(continues)

- $\sum_{i=1}^{n} X_i Y_i$, the sum of the cross product of X and Y.

To define the equation of the slope, b_1, as

$$b_1 = \frac{SSXY}{SSX}$$

in which

$$SSXY = \sum_{i=1}^{n}(X_i - \bar{X})(Y_i - \bar{Y}) = \sum_{i=1}^{n} X_i Y_i - \frac{\left(\sum_{i=1}^{n} X_i\right)\left(\sum_{i=1}^{n} Y_i\right)}{n}$$

and

$$SSX = \sum_{i=1}^{n}(X_i - \bar{X})^2 = \sum_{i=1}^{n} X_i^2 - \frac{\left(\sum_{i=1}^{n} X_i\right)^2}{n}$$

These equations, in turn, allow you to define the Y intercept as

$$b_0 = \bar{Y} - b_1 \bar{X}$$

For the moving company example, these sums and the sum of the squared Y values used on page 217 are calculated as follows:

Hours (*Y*)	Cu. Feet (*X*)	X^2	Y^2	*XY*
24	545	297,025	576.00	13,080.00
13.5	400	160,000	182.25	5,400.00
26.25	562	315,844	689.06	14,752.50
25	540	291,600	625.00	13,500.00
9	220	48,400	81.00	1,980.00
20	344	118,336	400.00	6,880.00
22	569	323,761	484.00	12,518.00
11.25	340	115,600	126.56	3,825.00
50	900	810,000	2,500.00	45,000.00
12	285	81,225	144.00	3,420.00
38.75	865	748,225	1,501.56	33,518.75
40	831	690,561	1,600.00	33.24
19.5	344	118,336	380.25	6,708.00
18	360	129,600	324.00	6,480.00
28	750	562,500	784.00	21,000.00
27	650	422,500	7290.00	17,550.00

Hours (Y)	Cu. Feet (X)	X^2	Y^2	XY
21	415	172,225	441.00	8,715.00
15	275	75,625	225.00	4,125.00
25	557	310,249	625.00	13,925.00
45	1,028	1,056,784	2,025.00	46,260.00
29	793	628,849	841.00	22,997.00
21	523	273,529	441.00	10,983.00
22	564	318,096	484.00	12,408.00
16.5	312	97,344	272.25	5,148.00
37	757	573,049	1,369.00	28,009.00
32	600	360,000	1,024.00	19,200.00
34	796	633,616	1,156.00	27,064.00
25	577	332,929	625.00	14,425.00
31	500	250,000	961.00	15,500.00
24	695	483,025	576.00	16,680.00
40	1,054	1,110,916	1,600.00	42,160.00
27	486	236,196	729.00	13,122.00
18	442	195,364	324.00	7,956.00
62.5	1,249	1,560,001	3,906.25	78,062.50
53.75	995	990,025	2,889.06	53,481.25
79.5	1,397	1,951,609	6,320.25	111,061.50
Sums: 1,042.5	22,520	16,842,944	37,960.50	790,134.50

Using these sums, you can compute the values of the slope b_1:

$$SSXY = \sum_{i=1}^{n}(X_i - \overline{X})(Y_i - \overline{Y}) = \sum_{i=1}^{n} X_i Y_i - \frac{\left(\sum_{i=1}^{n} X_i\right)\left(\sum_{i=1}^{n} Y_i\right)}{n}$$

$$SSXY = 790{,}134.5 - \frac{(22{,}520)(1{,}042.5)}{36}$$

$$= 790{,}134.5 - 652{,}141.66$$

$$= 137{,}992.84$$

$$SSX = \sum_{i=1}^{n}(X_i - \overline{X})^2 = \sum_{i=1}^{n} X_i^2 - \frac{\left(\sum_{i=1}^{n} X_i\right)^2}{n}$$

$$= 16{,}842{,}944 - \frac{(22{,}520)^2}{36}$$

$$= 16{,}842{,}944 - 14{,}087{,}511.11$$

$$= 2{,}755{,}432.89$$

(continues)

Because $b_1 = \frac{SSXY}{SSX}$

$$b_1 = \frac{137{,}992.84}{2{,}755{,}432.89}$$
$$= 0.05$$

With the value for the slope b_1, you can calculate the Y intercept as follows:

First, calculate the mean Y ($\overline{Y}$) and the mean X ($\overline{X}$) values:

$$\overline{Y} = \frac{\sum_{i=1}^{n} Y_i}{n} = \frac{1{,}042.5}{36} = 28.9583$$

$$\overline{X} = \frac{\sum_{i=1}^{n} X_i}{n} = \frac{22{,}520}{36} = 625.5555$$

Then use these results in the equation

$$b_0 = \overline{Y} - b_1 \overline{X}$$

$$b_0 = 28.9583 - (0.05)(625.5555)$$
$$= -2.3695$$

Regression Assumptions

important point

The assumptions necessary for performing a regression analysis are as follows:

- Normality of the variation around the line of regression
- Equality of variation in the Y values for all values of X
- Independence of the variation around the line of regression

The first assumption, normality, requires that the variation around the line of regression is normally distributed at each value of X. Like the t test and the ANOVA F test, regression analysis is fairly insensitive to departures from the normality assumption. As long as the distribution of the variation around the line of regression at each level of X is not extremely different from a normal distribution, inferences about the line of regression and the regression coefficients will not be seriously affected.

The second assumption, equality of variation, requires that the variation around the line of regression be constant for all values of *X*. This means that the variation is the same when *X* is a low value as when *X* is a high value. The equality of variation assumption is important for using the least-squares method of determining the regression coefficients. If there are serious departures from this assumption, other methods (see reference 5) can be used.

The third assumption, independence of the variation around the line of regression, requires that the variation around the regression line be independent for each value of *X*. This assumption is particularly important when data are collected over a period of time. In such situations, the variation around the line for a specific time period is often correlated with the variation of the previous time period.

Residual Analysis

The graphical method, **residual analysis**, enables you to evaluate whether the regression model that has been fitted to the data is an appropriate model and determine whether there are violations of the assumptions of the regression model.

Residual

CONCEPT The difference between the observed and predicted values of the dependent variable *Y* for a given value of *X*.

INTERPRETATION To evaluate the aptness of the fitted model, you plot the residuals on the vertical axis against the corresponding *X* values of the independent variable on the horizontal axis. If the fitted model is appropriate for the data, there will be no apparent pattern in this plot. However, if the fitted model is not appropriate, there will be a clear relationship between the *X* values and the residuals.

A residual plot for the moving company data fitted line of regression appears on the next page. In this figure, the cubic feet are plotted on the horizontal *X* axis and the residuals are plotted on the vertical *Y* axis. You see that although there is widespread scatter in the residual plot, no apparent pattern or relationship exists between the residuals and *X*. The residuals appear to be evenly spread above and below 0 for the differing values of *X*. This result enables you to conclude that the fitted straight-line model is appropriate for the moving company data.

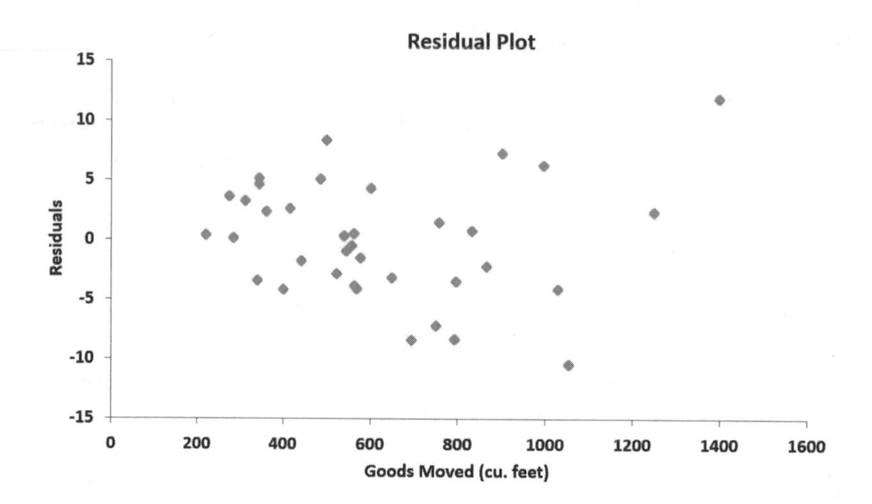

Evaluating the Assumptions

Different techniques, all involving the residuals, enable you to evaluate the regression assumptions.

For equality of variation, you use the same plot as you did to evaluate the aptness of the fitted model. For the moving company residual plot shown in the preceding figure, there do not appear to be major differences in the variability of the residuals for different X values. You can conclude that for this fitted model, there is no apparent violation in the assumption of equal variation at each level of X.

For the normality of the variation around the line of regression, you plot the residuals in a histogram (see Section 2.2), box-and-whisker plot (see Section 3.4), or a normal probability plot (see Section 5.4). From the histogram shown in the following figure for the moving company data, you can see that the data appear to be approximately normally distributed, with most of the residuals concentrated in the center of the distribution.

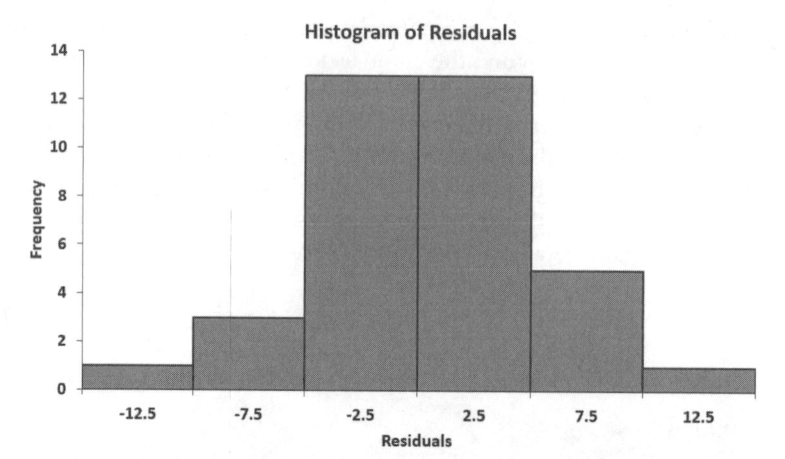

For the independence of the variation around the line of regression, you plot the residuals in the order or sequence in which the observed data was obtained, looking for a relationship between consecutive residuals. If you can see such a relationship, the assumption of independence is violated. Because these data were not collected over time, you do not need to evaluate this assumption.

10.3 Measures of Variation

After a regression model has been fit to a set of data, three measures of variation determine how much of the variation in the dependent variable *Y* can be explained by variation in the independent variable *X*.

Regression Sum of Squares (*SSR*)

CONCEPT The variation that is due to the relationship between *X* and *Y*.

INTERPRETATION The regression sum of squares (*SSR*) is equal to the sum of the squared differences between the *Y* values that are predicted from the regression equation and the mean value of *Y*:

$$SSR = \text{Sum (Predicted } Y \text{ value} - \text{Mean } Y \text{ value})^2$$

Error Sum of Squares (*SSE*)

CONCEPT The variation that is due to factors other than the relationship between *X* and *Y*.

INTERPRETATION The error sum of squares (*SSE*) is equal to the sum of the squared differences between each observed *Y* value and the predicted value of *Y*:

$$SSE = \text{Sum (Observed } Y \text{ value} - \text{Predicted } Y \text{ value})^2$$

Total Sum of Squares (*SST*)

CONCEPT The measure of variation of the Y_i values around their mean.

INTERPRETATION The total sum of squares (*SST*) is equal to the sum of the squared differences between each observed *Y* value and the mean value of *Y*:

$$SST = \text{Sum (Observed } Y \text{ value} - \text{Mean } Y \text{ value})^2$$

The total sum of squares is also equal to the sum of the regression sum of squares and the error sum of squares. For the WORKED-OUT PROBLEM of the previous section, *SSR* is 6,910.7189, *SSE* (called residual) is 860.7186, and *SST* is 7,771.4375. (Observe that 7,771.4375 is the sum of 6,910.7189 and 860.7186.)

You use symbols introduced earlier in this chapter to write the equations for the measures of variation in a regression analysis.

The equation for total sum of squares *SST* can be expressed in two ways:

$$SST = \sum_{i=1}^{n}(Y_i - \bar{Y})^2 \text{ equivalent to } \sum_{i=1}^{n} Y_i - \frac{\left(\sum_{i=1}^{n} Y_i\right)^2}{n}$$

or as

$$SST = SSR + SSE$$

The equation for the regression sum of squares (*SSR*) is

SSR = explained variation or regression sum of squares

$$SSR = \sum_{i=1}^{n}(\hat{Y}_i - \bar{Y})^2$$

which is equivalent to

$$= b_0 \sum_{i=1}^{n} Y_i + b_1 \sum_{i=1}^{n} X_i Y_i - \frac{\left(\sum_{i=1}^{n} Y_i\right)^2}{n}$$

The equation for the error sum of squares (*SSE*) is

SSE = unexplained variation or error sum of squares

$$= \sum_{i=1}^{n}(Y_i - \hat{Y}_i)^2$$

which is equivalent to

$$= \sum_{i=1}^{n} Y_i^2 - b_0 \sum_{i=1}^{n} Y_i - b_1 \sum_{i=1}^{n} X_i Y_i$$

For the moving company example on page 207,

$$SST = \text{total sum of squares} = \sum_{i=1}^{n}(Y_i - \bar{Y})^2 = \sum_{i=1}^{n} Y_i^2 - \frac{\left(\sum_{i=1}^{n} Y_i\right)^2}{n}$$

$$= 37{,}960.5 - \frac{(1{,}042.5)^2}{36}$$

$$= 37{,}960.5 - 30{,}189.0625$$

$$= 7{,}771.4375$$

$SSR =$ regression sum of squares

$$= \sum_{i=1}^{n} \hat{Y}_i - \bar{Y})^2$$

$$= b_0 \sum_{i=1}^{n} Y_i + b_1 \sum_{i=1}^{n} X_i Y_i - \frac{\left(\sum_{i=1}^{n} Y_i\right)^2}{n}$$

$$= (-2.3695)(1{,}042.5) + (0.05008)(790{,}134.5) - \frac{(1{,}042.5)^2}{36}$$

$$= 6{,}910.671$$

$SSE =$ error sum of squares

$$= \sum_{i=1}^{n}(Y_i - \hat{Y}_i)^2$$

$$= \sum_{i=1}^{n} Y_i^2 - b_0 \sum_{i=1}^{n} Y_i - b_1 \sum_{i=1}^{n} X_i Y_i$$

$$= 37{,}960.5 - (-2.3695)(1{,}042.5) - (0.05008)(790{,}134.5)$$

$$= 860.768$$

Calculated as $SSR + SSE$, the total sum of squares SST is 7,771.439, slightly different from the results from the first equation because of rounding errors.

The Coefficient of Determination

CONCEPT The ratio of the regression sum of squares to the total sum of squares, represented by the symbol r^2.

INTERPRETATION By themselves, *SSR*, *SSE*, and *SST* provide little that can be directly interpreted. The ratio of the regression sum of squares (*SSR*) to the total sum of squares (*SST*) measures the proportion of variation in *Y* that is explained by the independent variable *X* in the regression model. The ratio can be expressed as follows:

$$r^2 = \frac{\text{regression sum of squares}}{\text{total sum of squares}} = \frac{SSR}{SST}$$

For the moving company example, $SSR = 6{,}910.7189$ and $SST = 7{,}771.4375$ (see regression results on page 208). Therefore

$$r^2 = \frac{6{,}910.719}{7{,}771.4375} = 0.8892$$

This result for r^2 means that 89% of the variation in labor hours can be explained by the variability in the cubic footage to be moved. This shows a strong positive linear relationship between the two variables, because the use of a regression model has reduced the variability in predicting labor hours by 89%. Only 11% of the sample variability in labor hours can be explained by factors other than what is accounted for by the linear regression model that uses only cubic footage.

The Coefficient of Correlation

CONCEPT The measure of the strength of the linear relationship between two variables, represented by the symbol *r*.

INTERPRETATION The values of this coefficient vary from –1, which indicates perfect negative correlation, to +1, which indicates perfect positive correlation. The sign of the correlation coefficient *r* is the same as the sign of the slope in simple linear regression. If the slope is positive, *r* is positive. If the slope is negative, *r* is negative. The coefficient of correlation (*r*) is the square root of the coefficient of determination r^2.

For the moving company example, the coefficient of correlation, *r*, is +0.943, the positive (since the slope is positive) square root of 0.8892 (r^2). (Microsoft Excel labels the coefficient of correlation as "multiple r.") Because the coefficient is very close to +1.0, you can say that the relationship between cubic footage moved and labor hours is very strong. You can plausibly conclude that the increased volume that had to be moved is associated with increased labor hours.

In general, you must remember that just because two variables are strongly correlated, you cannot conclude that a cause-and-effect relationship exists between the variables.

Standard Error of the Estimate

CONCEPT The standard deviation around the fitted line of regression that measures the variability of the actual *Y* values from the predicted *Y*, represented by the symbol S_{YX}.

INTERPRETATION Although the least-squares method results in the line that fits the data with the minimum amount of variation, unless the coefficient of determination $r^2 = 1.0$, the regression equation is not a perfect predictor.

The variability around the line of regression was shown in the figure on page 208, which presented the scatter plot and the line of regression for the moving company data. You can see from that figure that some values are above the line of regression and other values are below the line of regression. For the moving company example, the standard error of the estimate (labeled as Standard Error in the figure on page 208) is equal to 5.03 hours.

Just as the standard deviation measures variability around the mean, the standard error of the estimate measures variability around the fitted line of regression. As you will see in Section 10.4, the standard error of the estimate can be used to determine whether a statistically significant relationship exists between the two variables.

equation blackboard (optional)

interested in math?

You use symbols introduced earlier in this chapter to write the equation for the standard error of the estimate:

$$S_{YX} = \sqrt{\frac{SSE}{n-2}} = \sqrt{\frac{\sum_{i=1}^{n}(Y_i - \hat{Y}_i)^2}{n-2}}$$

For the moving company problem, with *SSE* equal to 860.7186,

$$S_{YX} = \sqrt{\frac{860.7186}{36-2}}$$

$$S_{YX} = 5.0314$$

10.4 Inferences About the Slope

You can make inferences about the linear relationship between the variables in a population based on your sample results after using residual analysis to determine whether the assumptions of the least-squares regression model have not been seriously violated and that the straight-line model is appropriate.

t Test for the Slope

You can determine the existence of a significant relationship between the X and Y variables by testing whether β_1 (the population slope) is equal to 0. If this hypothesis is rejected, you conclude that evidence of a linear relationship exists. The null and alternative hypotheses are as follows:

H_0: $\beta_1 = 0$ (No linear relationship exists.)

H_1: $\beta_1 \neq 0$ (A linear relationship exists.)

The test statistic follows the t distribution with the degrees of freedom equal to the sample size minus 2. The test statistic is equal to the sample slope divided by the standard error of the slope:

$$t = \frac{\text{sample slope}}{\text{standard error of the slope}}$$

For the moving company example (see the results on page 208), the critical value of t at the level of significance of $\alpha = 0.05$ is 2.0322, the value of t is 16.52, and the p-value is 0.0000. (Microsoft Excel labels the t statistic "t Stat" on page 208.) Using the p-value approach, you reject H_0 because the p-value of 0.0000 is less than $\alpha = 0.05$. Using the critical value approach, you reject H_0 because $t = 16.52 > 2.0322$. You can conclude that a significant linear relationship exists between labor hours and the cubic footage moved.

You assemble symbols introduced earlier and the symbol for the standard error of the slope, S_{b_1}, to form the equation for the t statistic used in testing a hypothesis for a population slope β_1.

You begin by forming the equations for the standard error of the slope, S_{b_1} as

$$S_{b_1} = \frac{S_{YX}}{\sqrt{SSX}}$$

interested in math?

Then, you use the standard error of the slope, S_{b_1} to define the test statistic:

$$t = \frac{b_1 - \beta_1}{S_{b_1}}$$

The test statistic t follows a t distribution with $n - 2$ degrees of freedom.

For the moving company example, to test whether a significant relationship exists between the cubic footage and the labor hours at the level of significance $\alpha = 0.05$, refer to the calculation of SSX on page 210 and the standard error of the estimate on page 219:

$$S_{b_1} = \frac{S_{YX}}{\sqrt{SSX}}$$

$$= \frac{5.0314}{\sqrt{2{,}755{,}432.889}}$$

$$= 0.00303$$

Therefore, to test the existence of a linear relationship at the 0.05 level of significance, with

$$b_1 = +0.05008 \qquad n = 36 \qquad S_{b_1} = 0.00303$$

$$t = \frac{b_1 - \beta_1}{S_{b_1}}$$

$$= \frac{0.05008 - 0}{0.00303} = 16.52$$

Confidence Interval Estimate of the Slope (β_1)

You can also test the existence of a linear relationship between the variables by constructing a confidence interval estimate of β_1 and determining whether the hypothesized value ($\beta_1 = 0$) is included in the interval.

You construct the confidence interval estimate of the slope β_1 by multiplying the t statistic by the standard error of the slope and then adding and subtracting this product to the sample slope.

For the moving company example, the regression results on page 208 include the calculated lower and upper limits of the confidence interval estimate for the slope of cubic footage and labor hours. With 95% confidence, the lower limit is 0.0439 and the upper limit is 0.0562.

Because these values are above 0, you conclude that a significant linear relationship exists between labor hours and cubic footage moved. The confidence interval indicates that for each increase of 1 cubic foot moved, the mean labor hours are estimated to increase by at least 0.0439 hours but less than 0.0562 hours. Had the interval included 0, you would have concluded that no relationship exists between the variables.

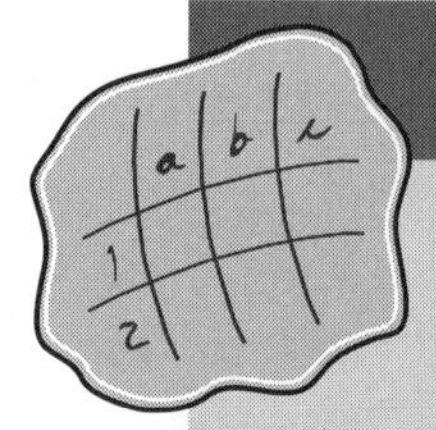

spreadsheet solution

Simple Linear Regression

Chapter 10 Simple Linear Regression contains the SLR spreadsheet that shows the WORKED-OUT PROBLEM 2 regression results for the moving company study data shown on page 207. Experiment with this spreadsheet by changing the confidence level of significance in cell L8.

Best Practices

Use the **LINEST(*cell range of Y variable, cell range of X variable*, True, True)** function to calculate the b_0 and b_1 coefficients and standard errors, r^2, the standard error of the estimate, the F test statistic, the residual degrees of freedom, and SSR and SSE.

Use the **T.INV.2T(*1 – confidence level, residual degrees of freedom*)** function to calculate the critical value for the t test.

Examine the **Chapter 10 Simple Linear Regression** RESIDUALS spreadsheet (not shown in Chapter 10) for a model for calculating residuals.

How-Tos

To calculate Significance F, use the **F.DIST.RT(*F critical value, regression degrees of freedom, residual degrees of freedom*)** function.

To calculate the *p*-values in the spreadsheet ANOVA table, use the **T.DIST.2T(*absolute value of the t test statistic, residual degrees of freedom*)** function.

Tip ATT6 in Appendix E describes using the Analysis ToolPak as a second way to perform a regression analysis.

Tip ADV6 in Appendix E explains more about how to use the **LINEST** function to calculate regression results.

Tip ADV7 in Appendix E explains how to modify **Chapter 10 Simple Linear Regression** for use with other data sets.

equation blackboard (optional)

interested in math?

You assemble symbols introduced earlier to form the equation for the confidence interval estimate of the slope β_1:

$$b_1 \pm t_{n-2} S_{b_1}$$

For the moving company example, b_1 has already been calculated on page 212 and the standard error of the slope S_{b_1} has already been calculated on page 221.

$b_1 = +0.05008 \quad n = 36 \quad S_{b_1} = 0.00303$

Thus, using 95% confidence, with degrees of freedom = 36 – 2 = 34,

$$b_1 \pm t_{n-2} S_{b_1}$$

$$= +0.05008 \pm (2.0322)(0.00303)$$

$$= +0.05008 \pm 0.0061$$

$$+0.0439 \le \beta_1 \le +0.0562$$

10.5 Common Mistakes Using Regression Analysis

Some of the common mistakes that people make when using regression analysis are as follows:

important point

- Lacking an awareness of the assumptions of least-squares regression
- Knowing how to evaluate the assumptions of least-squares regression
- Knowing what the alternatives to least-squares regression are if a particular assumption is violated
- Using a regression model without knowledge of the subject matter
- Predicting Y outside the relevant range of X

Most software regression analysis routines do not check for these mistakes. You must always use regression analysis wisely and always check that others who provide you with regression results have avoided these mistakes as well.

The following four sets of data illustrate some of the mistakes that you can make in a regression analysis.

Anscombe

Data Set A		Data Set B		Data Set C		Data Set D	
X_i	Y_i	X_i	Y_i	X_i	Y_i	X_i	Y_i
10	8.04	10	9.14	10	7.46	8	6.58
14	9.96	14	8.10	14	8.84	8	5.76
5	5.68	5	4.74	5	5.73	8	7.71
8	6.95	8	8.14	8	6.77	8	8.84
9	8.81	9	8.77	9	7.11	8	8.47
12	10.84	12	9.13	12	8.15	8	7.04
4	4.26	4	3.10	4	5.39	8	5.25
7	4.82	7	7.26	7	6.42	19	12.50
11	8.33	11	9.26	11	7.81	8	5.56
13	7.58	13	8.74	13	12.74	8	7.91
6	7.24	6	6.13	6	6.08	8	6.89

Source: Extracted from F. J. Anscombe, Graphs in Statistical Analysis, *American Statistician*, Vol. 27 (1973), pp. 17–21.

Anscombe (reference 1) showed that for the four data sets, the regression results are identical:

$$\text{predicted value of } Y = 3.0 + 0.5X_i$$
$$\text{standard error of the estimate} = 1.237$$
$$r^2 = 0.667$$
$$SSR = \text{regression sum of squares} = 27.51$$
$$SSE = \text{error sum of squares} = 13.76$$
$$SST = \text{total sum of squares} = 41.27$$

However, the four data sets are actually quite different, as scatter plots and residual plots for the four sets shown on the next page reveal.

By examining these visualizations, you discover how different these four data sets are:

- **Data set A:** The scatter plot seems to follow an approximate straight line, and the residual plot does not show any obvious patterns or outlying residuals.
- **Data set B:** The scatter plot suggests that a curvilinear regression model should be considered, and the residual plot reinforces that suggestion.

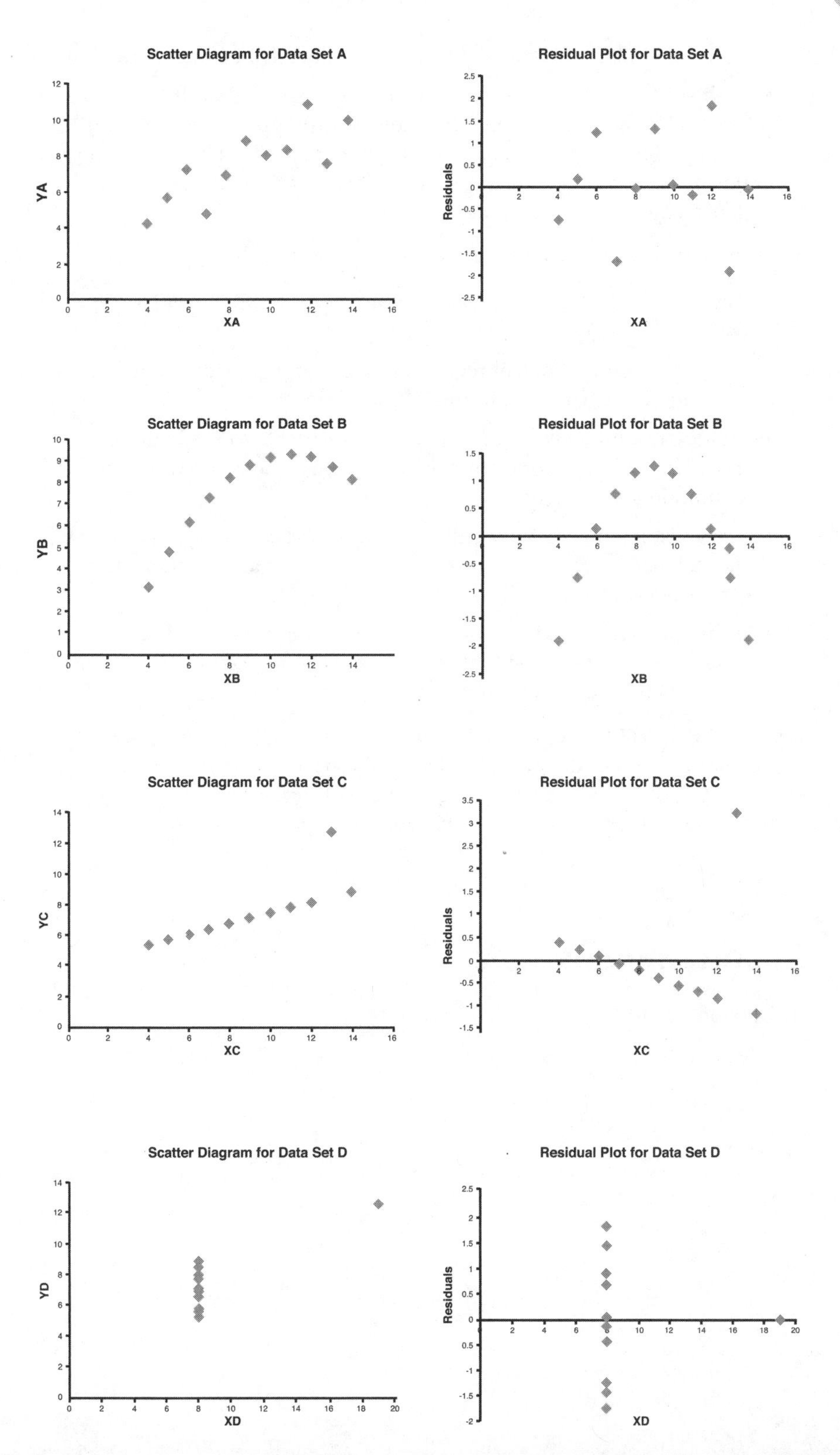

Scatter Diagram for Data Set A
YA
XA
Residual Plot for Data Set A
Residuals
XA
Scatter Diagram for Data Set B
YB
XB
Residual Plot for Data Set B
XB
Scatter Diagram for Data Set C
YC
XC
Residual Plot for Data Set C
Residuals
XC
Scatter Diagram for Data Set D
YD
XD
Residual Plot for Data Set D
Residuals
XD

- **Data set C**: Both the scatter and residual plots depict an extreme value.
- **Data set D**: The scatter plot reveals a single extreme value (at $X = 19$, $Y = 12.50$) that the residual plot shows greatly influences the fitted model. The regression model fit for these data should be evaluated cautiously because the regression coefficients for the model is heavily dependent on this single value.

To avoid the common mistakes of regression analysis, follow these steps:

- Always start with a scatter plot to observe the possible relationship between X and Y.
- Check the assumptions of regression after the regression model has been fit, before using the results of the model.
- Plot the residuals versus the independent variable to verify that the model fit is an appropriate model and to see if the equal variation assumption was violated.
- Use a histogram, box-and-whisker plot, or normal probability plot of the residuals to visually evaluate if the normality assumption has been seriously violated.
- If the evaluation of the residuals indicates violations in the assumptions, use alternative methods to least-squares regression or alternative least-squares models (see reference 5), as appropriate.

If the evaluation of the residuals does not indicate violations in the assumptions, then you can undertake the inferential aspects of the regression analysis. You can conduct a test for the significance of the slope and you can construct a confidence interval estimate of the slope.

Important Equations

Regression equation

$$(10.1) \quad \hat{Y}_i = b_0 + b_1 X_i$$

Slope

$$(10.2) \quad b_1 = \frac{SSXY}{SSX}$$

Sum of Squares

$$(10.3) \quad SSXY = \sum_{i=1}^{n}(X_i - \bar{X})(Y_i - \bar{Y}) = \sum_{i=1}^{n} X_i Y_i - \frac{\left(\sum_{i=1}^{n} X_i\right)\left(\sum_{i=1}^{n} Y_i\right)}{n}$$

and

$$SSX = \sum_{i=1}^{n}(X_i - \bar{X})^2 = \sum_{i=1}^{n} X_i^2 - \frac{\left(\sum_{i=1}^{n} X_i\right)^2}{n}$$

Y intercept

$$(10.4) \quad b_0 = \bar{Y} - b_1 \bar{X}$$

Total Sum of Squares

$$(10.5) \quad SST = \sum_{i=1}^{n}(Y_i - \bar{Y})^2 \text{ equivalent to } \sum_{i=1}^{n} Y_i - \frac{\left(\sum_{i=1}^{n} Y_i\right)^2}{n}$$

$$(10.6) \quad SST = SSR + SSE$$

Regression Sum of Squares

(10.7) SSR = explained variation or regression sum of squares

$$= \sum_{i=1}^{n}(\hat{Y}_i - \bar{Y})^2$$

which is equivalent to

$$= b_0 \sum_{i=1}^{n} Y_i + b_1 \sum_{i=1}^{n} X_i Y_i - \frac{\left(\sum_{i=1}^{n} Y_i\right)^2}{n}$$

Error Sum of Squares

(10.8) SSE = unexplained variation or error sum of squares

$$= \sum_{i=1}^{n}(Y_i - \hat{Y}_i)^2$$

which is equivalent to

$$= \sum_{i=1}^{n} Y_i^2 - b_0 \sum_{i=1}^{n} Y_i - b_1 \sum_{i=1}^{n} X_i Y_i$$

Coefficient of Determination

(10.9) $$r^2 = \frac{\text{regression sum of squares}}{\text{total sum of squares}} = \frac{SSR}{SST}$$

Coefficient of Correlation

(10.10) $$r = \sqrt{r^2}$$

If b_1 is positive, r is positive. If b_1 is negative, r is negative.

Standard Error of the Estimate

(10.11) $$S_{YX} = \sqrt{\frac{SSE}{n-2}} = \sqrt{\frac{\sum_{i=1}^{n}(Y_i - \hat{Y}_i)^2}{n-2}}$$

t Test for the Slope

(10.12) $$t = \frac{b_1 - \beta_1}{S_{b_1}}$$

One-Minute Summary

Simple Linear Regression

- Least-squares method
- Measures of variation
- Residual analysis
- *t* test for the significance of the slope
- Confidence interval estimate of the slope

Test Yourself

Short Answers

1. The Y intercept (b_0) represents the:
   (a) predicted value of Y when $X = 0$
   (b) change in Y per unit change in X
   (c) predicted value of Y
   (d) variation around the regression line

2. The slope (b_1) represents:
   (a) predicted value of Y when $X = 0$
   (b) change in Y per unit change in X
   (c) predicted value of Y
   (d) variation around the regression line

3. The standard error of the estimate is a measure of:
   (a) total variation of the Y variable
   (b) the variation around the regression line
   (c) explained variation
   (d) the variation of the X variable

4. The coefficient of determination (r^2) tells you:
   (a) that the coefficient of correlation (r) is larger than 1
   (b) whether the slope has any significance
   (c) whether the regression sum of squares is greater than the total sum of squares
   (d) the proportion of total variation that is explained

5. In performing a regression analysis involving two numerical variables, you assume:
   (a) the variances of X and Y are equal
   (b) the variation around the line of regression is the same for each X value
   (c) that X and Y are independent
   (d) All of the above

6. Which of the following assumptions concerning the distribution of the variation around the line of regression (the residuals) is correct?
   (a) The distribution is normal.
   (b) All of the variations are positive.
   (c) The variation increases as X increases.
   (d) Each residual is dependent on the previous residual.

7. The residuals represent:
   (a) the difference between the actual Y values and the mean of Y
   (b) the difference between the actual Y values and the predicted Y values
   (c) the square root of the slope
   (d) the predicted value of Y when $X = 0$

8. If the coefficient of determination (r^2) = 1.00, then:
   (a) the Y intercept must equal 0
   (b) the regression sum of squares (SSR) equals the error sum of squares (SSE)
   (c) the error sum of squares (SSE) equals 0
   (d) the regression sum of squares (SSR) equals 0

9. If the coefficient of correlation (r) = –1.00, then:
   (a) All the data points must fall exactly on a straight line with a slope that equals 1.00.
   (b) All the data points must fall exactly on a straight line with a negative slope.
   (c) All the data points must fall exactly on a straight line with a positive slope.
   (d) All the data points must fall exactly on a horizontal straight line with a zero slope.

10. Assuming a straight line (linear) relationship between X and Y, if the coefficient of correlation (r) equals –0.30:
    (a) there is no correlation
    (b) the slope is negative
    (c) variable X is larger than variable Y
    (d) the variance of X is negative

11. The strength of the linear relationship between two numerical variables is measured by the:
    (a) predicted value of Y
    (b) coefficient of determination
    (c) total sum of squares
    (d) Y intercept

12. In a simple linear regression model, the coefficient of correlation and the slope:
    (a) may have opposite signs
    (b) must have the same sign
    (c) must have opposite signs
    (d) are equal

Answer True or False:

13. The regression sum of squares (*SSR*) can never be greater than the total sum of squares (*SST*).
14. The coefficient of determination represents the ratio of *SSR* to *SST.*
15. Regression analysis is used for prediction, while correlation analysis is used to measure the strength of the association between two numerical variables.
16. The value of r is always positive.
17. When the coefficient of correlation $r = -1$, a perfect relationship exists between X and Y.
18. If no apparent pattern exists in the residual plot, the regression model fit is appropriate for the data.
19. If the range of the X variable is between 100 and 300, you should not make a prediction for $X = 400$.
20. If the *p*-value for a t test for the slope is 0.021, the results are significant at the 0.01 level of significance.

Fill in the blank:

21. The residual represents the difference between the observed value of Y and the ________ value of Y.
22. The change in Y per unit change in X is called the ________.
23. The ratio of the regression sum of squares (*SSR*) to the total sum of squares (*SST*) is called the ____________.
24. In simple linear regression, if the slope is positive, then the coefficient of correlation must also be _______.
25. One of the assumptions of regression is that the residuals around the line of regression follow the __________ distribution.

Answers to Test Yourself Short Answers

1. a
2. b
3. b
4. d
5. b
6. a
7. b
8. c
9. b
10. b
11. b
12. b
13. True
14. True
15. True
16. False
17. True
18. True

19. True
20. False
21. predicted
22. slope
23. coefficient of determination
24. positive
25. normal

Problems

1. The fair market value (in thousands of dollars) and property size (in acres) was collected for a sample of 30 single-family homes located in Glen Cove, New York (see the following table). Develop a simple linear regression model to predict appraised value based on the land area of the property.

GlenCove

Fair Market Value ($000)	Property Size (Acres)	Fair Market Value ($000)	Property Size (Acres)
522.9	0.2297	334.3	0.1714
425.0	0.2192	437.4	0.3849
539.2	0.1630	644.0	0.6545
628.2	0.4608	387.8	0.1722
490.4	0.2549	399.8	0.1435
487.7	0.2290	356.4	0.2755
370.3	0.1808	346.9	0.1148
777.9	0.5015	541.8	0.3636
347.1	0.2229	388.0	0.1474
756.8	0.1300	564.0	0.2281
389.0	0.1763	454.4	0.4626
889.0	1.3100	417.3	0.1889
452.2	0.2520	318.8	0.1228
412.4	0.1148	519.8	0.1492
338.3	0.1693	310.2	0.0852

(a) Assuming a linear relationship, use the least-squares method to compute the regression coefficients b_0 and b_1. State the regression equation for predicting the fair market value based on the land area.

(b) Interpret the meaning of the Y intercept b_0 and the slope b_1 in this problem.

(c) Explain why the regression coefficient, b_0, has no practical meaning in the context of this problem.

(d) Predict the fair market value for a house that has a property size of 0.25 acres.

(e) Compute the coefficient of determination, r^2, and interpret its meaning.

(f) Perform a residual analysis on the results and determine the adequacy of the model.

(g) Determine whether a significant relationship exists between fair market value and the property size at the 0.05 level of significance.

(h) Construct a 95% confidence interval estimate of the population slope between the fair market value and the property size.

2. Measuring the height of a California redwood tree is a very difficult undertaking because these trees grow to heights of more than 300 feet. People familiar with these trees understand that the height of a California redwood tree is related to other characteristics of the tree, including the diameter of the tree at the breast height of a person. The following data represent the height (in feet) and diameter at breast height of a person for a sample of 21 California redwood trees.

Redwood

Height	Diameter at breast height	Height	Diameter at breast height
122.0	20	164.0	40
193.5	36	203.3	52
166.5	18	174.0	30
82.0	10	159.0	22
133.5	21	205.0	42
156.0	29	223.5	45
172.5	51	195.0	54
81.0	11	232.5	39
148.0	26	190.5	36
113.0	12	100.0	8
84.0	13		

(a) Assuming a linear relationship, use the least-squares method to compute the regression coefficients b_0 and b_1. State the regression equation that predicts the height of a tree based on the tree's diameter at breast height of a person.

(b) Interpret the meaning of the slope in this equation.

(c) Predict the height for a tree that has a breast diameter of 25 inches.

(d) Interpret the meaning of the coefficient of determination in this problem.

(e) Perform a residual analysis on the results and determine the adequacy of the model.

(f) Determine whether a significant relationship exists between the height of redwood trees and the breast diameter at the 0.05 level of significance.

(g) Construct a 95% confidence interval estimate of the population slope between the height of the redwood trees and breast diameter.

3. A baseball analyst would like to study various team statistics a recent baseball season to determine which variables might be useful in predicting the number of wins achieved by teams during the season. He has decided to begin by using a team's earned run average (ERA), a measure of pitching performance, to predict the number of wins. The data for the 30 Major League Baseball teams are as follows:

(Hint: First, determine which are the independent and dependent variables.)

Baseball

(a) Assuming a linear relationship, use the least-squares method to compute the regression coefficients b_0 and b_1. State the regression equation for predicting the number of wins based on the ERA.

(b) Interpret the meaning of the *Y* intercept, b_0, and the slope, b_1, in this problem.

(c) Predict the number of wins for a team with an ERA of 3.75.

(d) Compute the coefficient of determination, r^2, and interpret its meaning.

(e) Perform a residual analysis on your results and determine the adequacy of the fit of the model.

(f) At the 0.05 level of significance, does evidence exist of a linear relationship between the number of wins and the ERA?

(g) Construct a 95% confidence interval estimate of the slope.

(h) What other independent variables might you include in the model?

Restaurants

4. Zagat's publishes restaurant ratings for various locations in the United States. For this analysis, a sample of 50 restaurants located in an urban area (New York City) and 50 restaurants located in a suburb of New York City are selected and the Zagat rating for food, decor, service, and the cost per person for each restaurant is recorded. Develop a regression model to predict the cost per person, based on a variable that represents the sum of the ratings for food, decor, and service.

Source: Extracted from *Zagat Survey 2013 New York City Restaurants* and *Zagat Survey 2012–2013, Long Island Restaurants*.

(a) Assuming a linear relationship, use the least-squares method to compute the regression coefficients b_0 and b_1. State the regression equation for predicting the cost per person based on the summated rating.

(b) Interpret the meaning of the Y intercept, b_0, and the slope, b_1, in this problem.

(c) Predict the cost per person for a restaurant with a summated rating of 50.

(d) Compute the coefficient of determination, r^2, and interpret its meaning.

(e) Perform a residual analysis on your results and determine the adequacy of the fit of the model.

(f) At the 0.05 level of significance, does evidence exist of a linear relationship between the cost per person and the summated rating?

(g) Construct a 95% confidence interval estimate of the population slope between the cost per person and the summated rating.

(h) How useful do you think the summated rating is as a predictor of cost? Explain.

Answers to Test Yourself Problems

1. (a) $b_0 = 354.9934$, $b_1 = 434.5435$; Predicted fair market value = 354.9934 + 434.5435 acre.

   (b) Each increase by one acre in property size is estimated to increase value by $434.5435 thousands.

   (c) The interpretation of b_0 has no practical meaning here because it would represent the estimated fair market value of a house that has no property size.

   (d) Predicted fair market value = 354.9934 + 434.5435 (0.25) = $463.63 thousands.

   (e) $r^2 = 0.5069$. 50.69% of the variation in fair market value of a house can be explained by variation in property size.

   (f) There is no particular pattern in the residual plot, and the model appears to be adequate.

   (g) $t = 5.365$; p-value $0.0000 < 0.05$ (or $t = 5.365 > 2.0484$). Reject H_0 at the 5% level of significance. There is evidence of a significant linear relationship between fair market value and property size.

   (h) $268.30 < \beta_1 < 600.457$

2. (a) $b_0 = 78.7963$, $b_1 = 2.6732$; Predicted height = 78.7963 + 2.6732 diameter of the tree at breast height of a person (in inches).

   (b) For each additional inch in the diameter of the tree at breast height of a person, the height of the tree is estimated to increase by 2.6732 feet.

(c) Predicted height = 78.7963 + 2.6732 (25) = 145.6267 feet

(d) $r^2 = 0.7288$. 72.88% of the total variation in the height of the tree can be explained by the variation of the diameter of the tree at breast height of a person.

(e) There is no particular pattern in the residual plot, and the model appears to be adequate.

(f) $t = 7.1455$; p-value = virtually $0 < 0.05$ ($t = 7.1455 > 2.093$). Reject H_0. There is evidence of a significant linear relationship between the height of the tree and the diameter of the tree at breast height of a person.

(g) $1.8902 < \beta_1 < 3.4562$

3. (a) $b_0 = 168.7804$, $b_1 = 22.7194$; Predicted wins = 168.7804 – 22.7194 ERA.

(b) For each additional earned run allowed, the number of wins is estimated to decrease by 22.7194.

(c) Predicted wins = 168.7804 – 22.7194 (3.75) = 83.5824.

(d) $r^2 = 0.5589$. 55.89% of the total variation in the number of wins can be explained by the variation of the ERA.

(e) There is no particular pattern in the residual plot, and the model appears to be adequate.

(f) $t = -5.9565$ p-value = virtually $0 < 0.05$ ($t = -5.9565 < -2.0484$). Reject H_0. There is evidence of a significant linear relationship between the number of wins and the ERA.

(g) $-30.5325 < \beta_1 < -14.9064$

(h) Among the independent variables you could consider including in the model are runs scored per game, hits allowed, saves, walks allowed, and errors. For a discussion of multiple regression models that consider several independent variables, see Chapter 11.

4. (a) $b_0 = -46.7718$, $b_1 = 1.4963$ predicted cost = –46.7718 + 1.4963 summated rating.

(b) For each additional unit increase in summated rating, the cost per person is estimated to increase by $1.50. Because no restaurant will receive a summated rating of 0, it is inappropriate to interpret the Y intercept.

(c) Predicted cost = –46.7718 + 1.4963 (50) = $28.04.

(d) $r^2 = 0.5458$. 54.58% of the variation in the cost per person can be explained by the variation in the summated rating.

(e) There is no obvious pattern in the residuals so the assumptions of regression are met. The model appears to be adequate.

(f) $t = 10.8524$, the p-value is virtually $0 < 0.05$ (or $t = 10.8524 > 1.9845$); reject H_0. There is evidence of a linear relationship between cost per person and summated rating.

(g) $1.2227 \le \beta_1 \le 1.7699$

(h) The linear regression model appears to have provided an adequate fit and shows a significant linear relationship between price per person and summated rating. Because 54.58% of the variation in the cost per person can be explained by the variation in summated rating, summated rating is moderately useful in predicting the cost per person.

References

1. Anscombe, F. J. "Graphs in Statistical Analysis." *American Statistician* 27 (1973): 17–21.
2. Berenson, M. L., D. M. Levine, and K. A. Szabat. *Basic Business Statistics: Concepts and Applications*, Thirteenth Edition. Upper Saddle River, NJ: Pearson Education, 2015.
3. Hosmer, D. W., and S. Lemeshow, *Applied Logistic Regression*, Third ed. (New York: Wiley, 2013).
4. Johnson, Stephen. "The Trouble with QSAR (or How I Learned to Stop Worrying and Embrace Fallacy." *Journal of Chemical Information and Modeling* 48 (2008): 25–26.
5. Kutner, M. H., C. Nachtsheim, J. Neter, and W. Li. *Applied Linear Statistical Models* 5th Ed. (New York: McGraw-Hill-Irwin, 2005).
6. Levine, D. M., D. Stephan, and K. A. Szabat. *Statistics for Managers Using Microsoft Excel*, Seventh Edition. Upper Saddle River, NJ: Pearson Education, 2014.
7. Microsoft Excel 2013. Redmond, WA: Microsoft Corporation, 2012.
8. Vidakovic, B. *Statistics for Bioengineering Sciences: With MATLAB and WinBUGS Support*. New York: Springer Science+Business Media, 2011.

CHAPTER 11

Multiple Regression

Chapter 10 discussed the simple linear regression model that uses one numerical independent variable X to predict the value of a numerical dependent variable Y. Often you can make better predictions if you use more than one independent variable. This chapter introduces you to multiple regression models that use two or more independent variables (Xs) to predict the value of a dependent variable (Y).

11.1 The Multiple Regression Model

CONCEPT The statistical method that extends the simple linear regression model on page 206 by assuming a straight-line or linear relationship between each independent variable and the dependent variable.

WORKED-OUT PROBLEM 1 In Chapter 10, when analyzing the moving company data, you used the cubic footage to be moved to predict the labor hours. In addition to cubic footage (X_1), now you are also going to consider the number of large pieces of furniture (such as beds, couches, china closets, and dressers) that need to be moved (X_2).

Moving

Hours	Cu. Feet Moved	Large Pieces	Hours	Cu. Feet Moved	Large Pieces
24.00	545	3	25.00	557	2
13.50	400	2	45.00	1,028	5
26.25	562	2	29.00	793	4
25.00	540	2	21.00	523	3
9.00	220	1	22.00	564	3
20.00	344	3	16.50	312	2
22.00	569	2	37.00	757	3
11.25	340	1	32.00	600	3
50.00	900	6	34.00	796	3
12.00	285	1	25.00	577	3
38.75	865	4	31.00	500	4
40.00	831	4	24.00	695	3
19.50	344	3	40.00	1,054	4
18.00	360	2	27.00	486	3
28.00	750	3	18.00	442	2
27.00	650	2	62.50	1,249	5
21.00	415	2	53.75	995	6
15.00	275	2	79.50	1,397	7

Spreadsheet results for these data are as follows:

	A	B	C	D	E	F	G	H	I
1	Multiple Regression Analysis for Moving Company Data								
2									
3	*Regression Statistics*								
4	Multiple R	0.96578							
5	R Square	0.93274							
6	Adjusted R Square	0.92866							
7	Standard Error	3.98000							
8	Observations	36							
9									
10	ANOVA								
11		*df*	*SS*	*MS*	*F*	*Significance F*			
12	Regression	2	7248.705601	3624.353	228.805	4.55335E-20			
13	Residual	33	522.7318989	15.84036					
14	Total	35	7771.4375						
15									
16		*Coefficients*	*Standard Error*	*t Stat*	*P-value*	*Lower 95%*	*Upper 95%*	*Lower 95.0%*	*Upper 95.0%*
17	Intercept	-3.9152	1.6738	-2.3391	0.0255	-7.3206	-0.5099	-7.3206	-0.5099
18	Cubic Feet Moved	0.0319	0.0046	6.9339	0.0000	0.0226	0.0413	0.0226	0.0413
19	Large Pieces	4.2228	0.9142	4.6192	0.0001	2.3629	6.0828	2.3629	6.0828

Net Regression Coefficients

CONCEPT The coefficients that measure the change in Y per unit change in a particular X, holding constant the effect of the other X variables. (Net regression coefficients are also known as **partial regression coefficients**.)

INTERPRETATION In the simple regression model in Chapter 10, the slope represents the change in Y per unit change in X and considers only the single independent variable included in the model. In the multiple regression model with two independent variables, there are two net regression coefficients: b_1, the slope of Y with X_1 represents the change in Y per unit change in X_1, taking into account the effect of X_2; and b_2, the slope of Y with X_2 represents the change in Y per unit change in X_2, taking into account the effect of X_1.

For example, the results for the moving company study show that the slope of Y with X_1, b_1 is 0.0319 and the slope of Y with X_2, b_2, is 4.2228. The slope b_1 means that for each increase of 1 unit in X_1, the value of Y is estimated to increase by 0.0319 units, holding constant the effect of X_2. In other words, holding constant the number of pieces of large furniture, for each increase of 1 cubic foot in the amount to be moved, the fitted model predicts that the labor hours are estimated to increase by 0.0319 hours.

The slope b_2 (+4.2228) means that for each increase of 1 unit in X_2, the value of Y is estimated to increase by 4.2228 units, holding constant the effect of X_1. In other words, holding constant the amount to be moved, for each additional piece of large furniture, the fitted model predicts that the labor hours are estimated to increase by 4.2228 hours.

Another way to interpret this "net effect" is to think of two moves with an equal number of pieces of large furniture. If the first move consists of 1 cubic foot more than the other move, the "net effect" of this difference is that the first move is predicted to take 0.0319 more labor hours than the other move. To interpret the net effect of the number of pieces of large furniture, you can consider two moves that have the same cubic footage. If the first move has one additional piece of large furniture, the net effect of this difference is that the first move is predicted to take 4.2228 more labor hours than the other move.

Adding the Y intercept, b_0, to the net regression coefficients b_1 and b_2 creates the multiple regression equation, which for a model with two independent variables is:

$$\text{Predicted } Y = b_0 + b_1X_1 + b_2X_2$$

For example, the results for the moving company study show that the *Y* intercept, b_0, is –3.915. This means that for this study the multiple regression equation is

$$\text{Predicted value of labor hours} = -3.915 + (0.0319 \times \text{cubic feet moved}) + (4.2228 \times \text{large furniture})$$

(Recall from Chapter 10 that the *Y* intercept represents the estimated value of *Y* when *X* equals 0. In this example, because the cubic feet moved cannot be less than 0, the *Y* intercept has no practical interpretation. Recall, also, that a regression model is only valid within the ranges of the independent variables.)

Predicting the Dependent Variable *Y*

As in simple linear regression, you can use the multiple regression equation to predict values of the dependent variable.

WORKED-OUT PROBLEM 2 Using the multiple regression model developed in **WORKED-OUT PROBLEM 1**, you want to predict the labor hours for a move with 500 cubic feet with three large pieces of furniture to be moved. You predict that the labor hours for such a move are 24.715 (–3.915 + 0.0319 × 500 + 4.2228 × 3).

11.2 Coefficient of Multiple Determination

CONCEPT The statistic that represents the proportion of the variation in *Y* that is explained by the set of independent variables included in the multiple regression model.

INTERPRETATION The coefficient of multiple determination is analogous to the coefficient of determination (r^2) that measures the variation in *Y* that is explained by the independent variable *X* in the simple linear regression model (see Section 10.3).

WORKED-OUT PROBLEM 3 You need to calculate this coefficient for the moving company study. In the multiple regression results (see page 240), the ANOVA summary table shows that *SSR* is 7,248.71 and *SST* is 7,771.44. Therefore,

$$r^2 = \frac{\text{regression sum of squares}}{\text{total sum of squares}} = \frac{SSR}{SST}$$

$$= \frac{7{,}248.71}{7{,}771.44}$$

$$= 0.9327$$

The coefficient of multiple determination, ($r^2 = 0.9327$), indicates that 93.27% of the variation in labor hours is explained by the variation in the cubic footage and the variation in the number of pieces of large furniture to be moved.

11.3 The Overall *F* Test

CONCEPT The test for the significance of the overall multiple regression model.

INTERPRETATION You use this test to determine whether a significant relationship exists between the dependent variable and the entire set of independent variables. Because there is more than one independent variable, you have the following null and alternative hypotheses:

> H_0: No linear relationship exists between the dependent variable and the independent variables.
>
> H_1: A linear relationship exists between the dependent variable and at least one of the independent variables.

The ANOVA summary table for the overall F test is as follows (n = sample size and k = number of independent variables):

Source	Degrees of Freedom	Sum of Squares	Mean Square (Variance)	F
Regression	k	SSR	$MSR = \frac{SSR}{k}$	$F = \frac{MSR}{MSE}$
Error	$n - k - 1$	SSE	$MSE = \frac{SSE}{n-k-1}$	
Total	$n - 1$	SST		

WORKED-OUT PROBLEM 4 You need to perform this test for the moving company study. In the multiple regression results (see page 240), the ANOVA summary table shows that the F statistic is 228.805 and the p-value = 0.000 (shown in the results as 4.55E-20). Because the p-value = 0.000 < 0.05, you reject H_0 and conclude that at least one of the independent variables (cubic footage and/or the number of pieces of large furniture moved) is related to labor hours.

11.4 Residual Analysis for the Multiple Regression Model

In Section 10.2, you used residual analysis to evaluate whether the simple linear regression model was appropriate for a set of data. For the multiple regression model with two independent variables, you construct and analyze these residual plots:

- Residuals versus the predicted value of Y
- Residuals versus the first independent variable X_1
- Residuals versus the second independent variable X_2
- Residuals versus time (if the data has been collected in time order)

If the residuals versus the predicted value of Y show a pattern for different predicted values of Y, there is evidence of a possible curvilinear effect in at least one independent variable, a possible violation to the assumption of equal variance, and/or the need to transform the Y variable. Patterns in the plot of the residuals versus an independent variable (the second and third type of plots) can indicate the existence of a curvilinear effect and, therefore, indicate the need to add a curvilinear independent variable to the multiple regression model (see references 1 and 2). Patterns in the residuals in the fourth type of plot can help determine whether the independence assumption has been violated when the data are collected in time order.

WORKED-OUT PROBLEM 5 You need to perform a residual analysis for the multiple regression model for the moving company study. The residual plots for this model are shown below.

In these plots, you see very little or no pattern in the relationship between the residuals and the predicted value of Y, the cubic feet moved (X_1), or the number of pieces of large furniture moved (X_2). You conclude that the multiple regression model is appropriate for predicting labor hours.

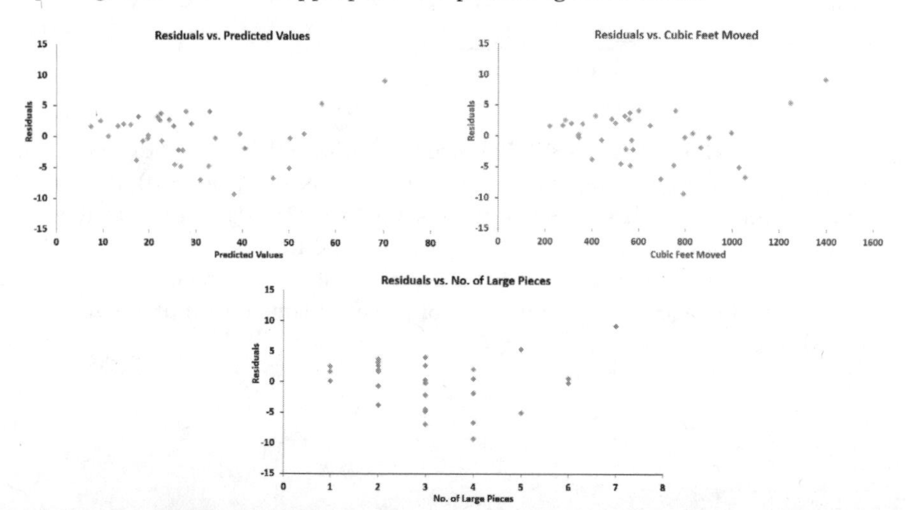

11.5 Inferences Concerning the Population Regression Coefficients

In Section 10.4, you tested the existence of the slope in a simple linear regression model to determine the significance of the relationship between X and Y. In addition, you constructed a confidence interval estimate of the population slope. In this section, these procedures are extended to the multiple regression model.

Tests of Hypothesis

As with the simple linear regression model, you use a t test to test a hypothesis concerning the population slope: For a multiple regression model with two independent variables, the null hypothesis for each independent variable is that no linear relationship exists between labor hours and the independent variable holding constant the effect of the other independent variables. The alternative hypothesis is that a linear relationship exists between labor hours and the independent variable holding constant the effect of the other independent variables. The t test of significance for a particular regression coefficient is actually a test for the significance of adding a particular variable into a regression model given that the other variable is included.

WORKED-OUT PROBLEM 6 You need to test the hypothesis concerning the population slopes of the multiple regression model for the moving company study. From the results on page 240, for the cubic feet moved independent variable, the t statistic is 6.93 and the p-value is 0.000. Because the p-value is $0.000 < 0.05$, you reject the null hypothesis and conclude that a linear relationship exists between labor hours and the cubic feet moved (X_1). For the number of pieces of large furniture moved independent variable, the t statistic is 4.62 and the p-value is 0.000. Because the p-value is $0.000 < 0.05$, you reject the null hypothesis and conclude that a linear relationship exists between labor hours and the number of pieces of large furniture moved (X_2). You conclude that because each of the two independent variables is significant, both should be included in the regression model.

Confidence Interval Estimation

As you did in simple linear regression (see Section 10.4), you can construct confidence interval estimates of the slope. You calculate the confidence interval estimate of the population slope by multiplying the t statistic by the standard error of the slope and then adding and subtracting this product to the sample slope.

WORKED-OUT PROBLEM 7 You want to determine the confidence interval estimates of the slope from the moving company study multiple regression model. From the results on page 240, with 95% confidence, the lower limit for the slope of the number of feet moved with labor hours is 0.0226 hours and the upper limit is 0.0413 hours. The confidence interval indicates that for each increase of 1 cubic foot moved, labor hours are estimated to increase by at least 0.0226 hours but less than 0.0413 hours, holding constant the number of pieces of large furniture moved. With 95% confidence, the lower limit of the slope of the number of pieces of large furniture moved and labor hours is 2.3629 hours and the upper limit is 6.0828 hours. The confidence interval indicates that for each increase of one piece of large furniture moved, labor hours are estimated to increase by at least 2.3629 hours but less than 6.0827 hours holding constant the cubic footage moved.

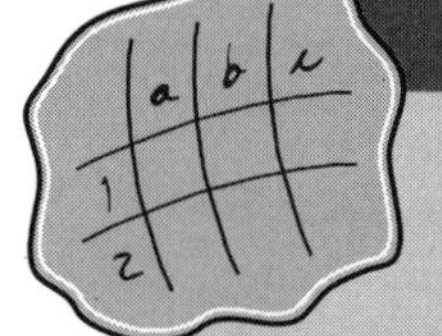

spreadsheet solution

Multiple Regression

Chapter 11 Multiple Regression contains the spreadsheet that shows the WORKED-OUT PROBLEM 1 regression results for the moving company study data shown on page 240. Experiment with this spreadsheet by changing the confidence level in cell L8.

Best Practices

Use the **LINEST**(***cell range of Y variable, cell range of X variable, True, True***) function to calculate the b_0 and b_1 coefficients and standard errors, r^2, the standard error of the estimate, the F test statistic, the residual degrees of freedom, and SSR and SSE.

Use the **T.INV.2T**(***1 – confidence level, residual degrees of freedom***) function to calculate the critical value for the t test.

Examine the **Chapter 10 Simple Linear Regression** RESIDUALS spreadsheet (not shown in Chapter 10) for a model for calculating residuals.

How-Tos

To calculate Significance F, use the **F.DIST.RT**(***F critical value, regression degrees of freedom, residual degrees of freedom***) function.

To calculate the *p*-values in the spreadsheet ANOVA table, use the **T.DIST.2T**(***absolute value of the t test statistic, residual degrees of freedom***) function.

Tip ATT6 in Appendix E describes using the Analysis ToolPak as a second way to perform a regression analysis.

Tip ADV6 in Appendix E explains more about how to use the LINEST function to calculate regression results.

One-Minute Summary

Multiple regression

- Use several independent variables to predict a dependent variable
- Net regression coefficients
- Coefficient of Multiple Determination
- Overall *F* Test
- Residual analysis
- *t* test for the significance of each independent variable

Test Yourself

Short Answers

1. In a multiple regression model involving two independent variables, if b_1 is +3.0, it means that:
   (a) The relationship between X_1 and Y is significant.
   (b) The estimated value of Y increases by 3 units for each increase of 1 unit of X_1, holding X_2 constant.
   (c) The estimated value of Y increases by 3 units for each increase of 1 unit of X_1, without regard to X_2.
   (d) The estimated value of Y is 3 when X_1 equals zero.
2. The coefficient of multiple determination
   (a) measures the variation around the predicted regression equation
   (b) measures the proportion of variation in Y that is explained by X_1 and X_2
   (c) measures the proportion of variation in Y that is explained by X_1 holding X_2 constant
   (d) will have the same sign as b_1
3. In a multiple regression model, the value of the coefficient of multiple determination
   (a) is between −1 and +1
   (b) is between 0 and +1
   (c) is between −1 and 0
   (d) can be any number

Answer True or False:

4. The interpretation of the slope is different in a multiple linear regression model as compared to a simple linear regression model.

5. The interpretation of the Y intercept is different in a multiple linear regression model as compared to a simple linear regression model.
6. In a multiple regression model with two independent variables, the coefficient of multiple determination measures the proportion of variation in Y that is explained by X_1 and X_2.
7. The slopes in a multiple regression model are called net regression coefficients.
8. The coefficient of multiple determination is calculated by taking the ratio of the regression sum of squares over the total sum of squares (SSR/SST) and subtracting that value from 1.
9. You have just developed a multiple regression model in which the value of coefficient of multiple determination is 0.35. To determine whether this indicates that the independent variables explain a significant portion of the variation in the dependent variable, you would perform an F-test.
10. From the coefficient of multiple determination, you cannot detect the strength of the relationship between Y and any individual independent variable.

Answers to Test Yourself Short Answers

1. b
2. b
3. b
4. True
5. False
6. True
7. True
8. False
9. True
10. True

Problems

1. The fair market value (in thousands of dollars), property size (in acres), and age of house (in years) was collected for a sample of 30 single-family homes located in Glen Cove, New York (see the following table). Develop a multiple linear regression model to predict the fair market value based on the property size and age, in years.

GlenCove

Fair Market Value ($000)	Property Size (Acres)	Age	Fair Market Value ($000)	Property Size (Acres)	Age
522.9	0.2297	56	334.3	0.1714	62
425.0	0.2192	61	437.4	0.3849	54
539.2	0.1630	39	644.0	0.6545	56
628.2	0.4608	28	387.8	0.1722	62
490.4	0.2549	56	399.8	0.1435	88
487.7	0.2290	98	356.4	0.2755	81
370.3	0.1808	58	346.9	0.1148	107
777.9	0.5015	17	541.8	0.3636	55
347.1	0.2229	62	388.0	0.1474	51
756.8	0.1300	25	564.0	0.2281	50
389.0	0.1763	64	454.4	0.4626	92
889.0	1.3100	62	417.3	0.1889	64
452.2	0.2520	56	318.8	0.1228	54
412.4	0.1148	22	519.8	0.1492	44
338.3	0.1693	74	310.2	0.0852	104

(a) State the multiple regression equation.

(b) Interpret the meaning of the slopes, b_1 and b_2, in this problem.

(c) Explain why the regression coefficient, b_0, has no practical meaning in the context of this problem.

(d) Predict the fair market value for a house that has a property size of 0.25 acres and is 55 years old.

(e) Compute the coefficient of multiple determination, r^2, and interpret its meaning.

(f) Perform a residual analysis on the results and determine the adequacy of the model.

(g) Determine whether a significant relationship exists between the fair market value and the two independent variables (property size and age of a house) at the 0.05 level of significance.

(h) At the 0.05 level of significance, determine whether each independent variable makes a significant contribution to the regression model. On the basis of these results, indicate the independent variables to include in this model.

(i) Construct a 95% confidence interval estimate of the population slope between fair market value and property size and between fair market value and age.

2. Measuring the height of a California redwood tree is a very difficult undertaking because these trees grow to heights of more than 300 feet. People familiar with these trees understand that the height of a California redwood tree is related to other characteristics of the tree, including the diameter of the tree at the breast height of a person and the thickness of the bark of the tree. The following data represent the height (in feet), diameter at breast height of a person, and bark thickness for a sample of 21 California redwood trees:

Redwood

Height	Diameter at Breast Height	Bark Thickness	Height	Diameter at Breast Height	Bark Thickness
122.0	20	1.1	164.0	40	2.3
193.5	36	2.8	203.3	52	2.0
166.5	18	2.0	174.0	30	2.5
82.0	10	1.2	159.0	22	3.0
133.5	21	2.0	205.0	42	2.6
156.0	29	1.4	223.5	45	4.3
172.5	51	1.8	195.0	54	4.0
81.0	11	1.1	232.5	39	2.2
148.0	26	2.5	190.5	36	3.5
113.0	12	1.5	100.0	8	1.4
84.0	13	1.4			

(a) State the multiple regression equation that predicts the height of a tree based on the tree's diameter at breast height and the thickness of the bark.

(b) Interpret the meaning of the slopes in this equation.

(c) Predict the height for a tree that has a breast diameter of 25 inches and a bark thickness of 2 inches.

(d) Interpret the meaning of the coefficient of multiple determination in this problem.

(e) Perform a residual analysis on the results and determine the adequacy of the model.

(f) Determine whether a significant relationship exists between the height of redwood trees and the two independent variables (breast diameter and the bark thickness) at the 0.05 level of significance.

(g) At the 0.05 level of significance, determine whether each independent variable makes a significant contribution to the regression model. Indicate the independent variables to include in this model.

(h) Construct a 95% confidence interval estimate of the population slope between the height of the redwood trees and breast diameter and between the height of redwood trees and the bark thickness.

3. A baseball analytics specialist wants to determine which variables are important in predicting a team's wins in a given season. He has collected data related to wins, earned run average (ERA), and runs scored per game for a recent season. Develop a model to predict the number of wins based on ERA and runs scored per game.

Baseball

Team	ERA	Runs Scored per Game	Wins
Baltimore	4.20	4.60	85
Boston	3.79	5.27	97
Chi. White Sox	3.98	3.69	63
Cleveland	3.82	4.60	92
Detroit	3.61	4.91	93
Houston	4.79	3.77	51
Kansas City	3.45	4.00	86
LA Angels	4.23	4.52	78
Minnesota	4.55	3.79	66
NY Yankees	3.94	4.01	85
Oakland	3.56	4.73	96
Seattle	4.31	3.85	71
Tampa Bay	3.75	4.29	91
Texas	3.62	4.49	91
Toronto	4.25	4.40	74
Arizona	3.92	4.23	81
Atlanta	3.18	4.25	96
Chi. Cubs	4.00	3.72	66
Cincinnati	3.38	4.31	90
Colorado	4.44	4.36	74
LA Dodgers	3.25	4.01	92
Miami	3.71	3.17	62
Milwaukee	3.84	3.95	74

Team	ERA	Runs Scored per Game	Wins
NY Mets	3.77	3.82	74
Philadelphia	4.32	3.77	73
Pittsburgh	3.26	3.91	94
St. Louis	3.42	4.83	97
San Diego	3.98	3.81	76
San Francisco	4.00	3.88	76
Washington	3.59	4.05	86

(a) State the multiple regression equation.

(b) Interpret the meaning of the slopes in this equation.

(c) Predict the number of wins for a team that has an ERA of 3.75 and has runs scored per game of 4.00.

(d) Perform a residual analysis on the results and determine whether the regression assumptions are valid.

(e) Is there a significant relationship between number of wins and the two independent variables (ERA and runs scored) at the 0.05 level of significance?

(f) Interpret the meaning of the coefficient of multiple determination.

(g) At the 0.05 level of significance, determine whether each independent variable makes a significant contribution to the regression model. Indicate the most appropriate regression model for this set of data.

(h) Construct 95% confidence interval estimates of the population slope between wins and ERA and between the number of wins and runs scored per game.

Auto

4. A consumer organization wants to develop a regression model to predict gasoline mileage (as measured by miles per gallon) based on the horsepower of the car's engine and the weight of the car (in pounds). A sample of 50 recent car models was selected (see the Auto file).
   (a) State the multiple regression equation.
   (b) Interpret the meaning of the slopes, b_1 and b_2, in this problem.
   (c) Explain why the regression coefficient, b_0, has no practical meaning in the context of this problem.
   (d) Predict the miles per gallon for cars that have 60 horsepower and weigh 2,000 pounds.
   (e) Perform a residual analysis on your results and determine the adequacy of the fit of the model.

(f) Compute the coefficient of multiple determination, r^2, and interpret its meaning.

(g) Determine whether a significant relationship exists between gasoline mileage and the two independent variables (horsepower and weight) at the 0.05 level of significance.

(h) At the 0.05 level of significance, determine whether each independent variable makes a significant contribution to the regression model. On the basis of these results, indicate the independent variables to include in this model.

(i) Construct a 95% confidence interval estimate of the population slope between gasoline mileage and horsepower and between gasoline mileage and weight.

Answers to Problems

1. (a) Predicted fair market value = 532.2883 + 407.1346 property size – 2.8257 age.

   (b) For a given age, each increase by one acre in property size is estimated to result in an increase in fair market value by $407.13 thousands. For a given property size, each increase of one year in age is estimated to result in the decrease in fair market value by $2.83 thousands.

   (c) The interpretation of b_0 has no practical meaning here because it would represent the estimated fair market value of a new house that has no property size.

   (d) Predicted fair market value = 532.2883 + 407.1346(0.25) – 2.8257(55) = $478.66 thousands.

   (e) $r^2 = 0.6988$; so 69.88% of the variation in fair market value of a house can be explained by variation in property size and age of the house.

   (f) There is no particular pattern in the residual plots, and the model appears to be adequate.

   (g) $F = 31.32$; *p*-value is virtually zero. Reject H_0 at 5% level of significance. Evidence exists of a significant linear relationship between fair market value and the two explanatory variables.

   (h) For property size: $t = 6.2827 > 2.0518$ or $p\text{-value} = 0.0000 < 0.05$ Reject H_0. Property size makes a significant contribution to the regression model after age is included. For age: $t = -4.1475 < -2.0518$ or $p\text{-value} = 0.0003 < 0.05$ Reject H_0. Age makes a significant contribution to the regression model after property size is included. Therefore, both property size and age should be included in the model.

   (i) $274.1702 < \beta_1 < 540.0990$; $-4.2236 < \beta_2 < -1.4278$

2. (a) Predicted height = 62.1411 + 2.0567 diameter of the tree at breast height of a person (in inches) + 15.6418 thickness of the bark (in inches).

   (b) Holding constant the effects of the thickness of the bark, for each additional inch in the diameter of the tree at breast height of a person, the height of the tree is estimated to increase by 2.0567 feet. Holding constant the effects of the diameter of the tree at breast height of a person, for each additional inch in the thickness of the bark, the height of the tree is estimated to increase by 15.6418 feet.

   (c) Predicted height = 62.1411 + 2.0567 (25) + 15.6418 (2) = 144.84 feet.

   (d) $r^2 = 0.7858$. 78.58% of the total variation in the height of the tree can be explained by the variation in the diameter of the tree at breast height of a person and the thickness of the bark of the tree.

   (e) The plot of the residuals against bark thickness indicates a potential pattern that might require the addition of curvilinear terms. One value appears to be an outlier in both plots.

   (f) $F = 33.0134$ with 2 and 18 degrees of freedom. p-value = virtually $0 < 0.05$. Reject H_0. At least one of the independent variables is linearly related to the dependent variable.

   (g) Breast height diameter: $t = 4.6448 > 2.1009$ or p-value $= 0.0002 < 0.05$ Reject H_0. Breast height diameter makes a significant contribution to the regression model after bark thickness is included. Bark thickness: $t = 2.1882 > 2.1009$ or p-value $= 0.0421 < 0.05$ Reject H_0. Bark thickness makes a significant contribution to the regression model after breast height diameter is included. Therefore, both breast height diameter and bark thickness should be included in the model.

   (h) $1.1264 \le \beta_1 \le 2.9870$; $0.6238 \le \beta_2 \le 30.6598$

3. (a) Predicted wins =87.7213 –18.9527 ERA +15.9626 runs scored per game.

   (b) For a given runs scored per game, for each increase of 1 in the ERA, the number of wins is estimated to decrease by 18.9527. For a given ERA, for each increase of one in the number of runs scored per game, the number of wins are estimated to increase by 15.9626.

   (c) Predicted wins = 87.7213 –18.9527 (3.75) +15.9626 (4) = 80.50.

   (d) There is no evidence of a pattern in the residual plot of ERA or runs scored per game.

   (e) $F = MSR/MSE = 1{,}914.3312/18.7162 = 102.282$; p-value $= 0.0000 < 0.05$. Reject H_0 at 5% level of significance. Evidence of a significant linear relationship exists between number of wins and the two explanatory variables.

(f) $r^2 = SSR/SST$ = 3,828.6625/4,334.0 = 0.8834; 88.34% of the variation in number of wins can be explained by variation in ERA and runs scored per game.

(g) For X_1: $t_{STAT} = b_1 / S_{b_1} = -9.2734 > -2.0518$ and p-value 0.0000 < 0.05, reject H_0. There is evidence that the variable X_1 contributes to a model already containing X_2. For X_2: $t_{STAT} = b_2 / S_{b_2} = 8.6683 > 2.0518$ and p-value 0.0000 < 0.05, reject H_0. Both variables X_1 and X_2 should be included in the model.

(h) $-23.1462 < \beta_1 < -14.7592$ $\quad 12.1842 < \beta_2 < 19.7411$

4. (a) Predicted miles per gallon (MPG) = 58.15708 – 0.11753 horsepower – 0.00687 weight.

(b) For a given weight, each increase of one unit in horsepower is estimated to result in a decrease in MPG of 0.11753. For a given horsepower, each increase of one pound is estimated to result in the decrease in MPG of 0.00687.

(c) The interpretation of b_0 has no practical meaning here because it would have involved estimating gasoline mileage when a car has 0 horsepower and 0 weight.

(d) Predicted miles per gallon (MPG) = 58.15708 – 0.11753 (60) – 0.00687 (2,000) = 37.365 MPG.

(e) There appears to be a curvilinear relationship in the plot of the residuals against the predicted values of MPG and the horsepower. Thus, curvilinear terms for the independent variables should be considered for inclusion in the model.

(f) F = 1,225.98685/17.444 = 70.2813; p-value = 0.0000 < 0.05 Reject H_0. Evidence of a significant linear relationship exists between MPG and horsepower and/or weight.

(g) r^2 = 2,451.974/3,271.842 = 0.7494 74.94% of the variation in MPG can be explained by variation in horsepower and variation in weight.

(h) For horsepower: $t < -3.60 < -2.0117$ or p-value = 0.0008 < 0.05 Reject H_0. There is evidence that horsepower contributes to a model already containing weight. For weight: $t < -4.9035 < -2.0117$ or p-value = 0.0000 < 0.05 Reject H_0. There is evidence that weight contributes to a model already containing horsepower. Both variables X_1 and X_2 should be included in the model.

(i) $-0.1832 < \beta_1 < -0.0519$ $\quad -0.0097 < \beta_2 < -0.00405$

References

1. Berenson, M. L., D. M. Levine, and K. A. Szabat. *Basic Business Statistics: Concepts and Applications*, 13th Ed. (Upper Saddle River, NJ: Pearson Education, 2015).
2. Kutner, M. H., C. Nachtsheim, J. Neter, and W. Li. *Applied Linear Statistical Models*, 5th Ed. (New York: McGraw-Hill-Irwin, 2005).
3. Microsoft Excel 2013. Redmond, WA: Microsoft Corp., 2012.

CHAPTER

12

Fundamentals of Analytics

Previous chapters discussed descriptive methods that present and summarize data and inferential methods used for confidence interval estimation, hypothesis testing, and regression. Analytics are methods that present, summarize, or infer something from data, too, but in ways that sometimes invert the methodology presented in earlier chapters. This chapter introduces you to the fundamental concepts of analytics.

12.1 Basic Vocabulary of Analytics

To best understand how analytics methods differ from methods discussed in earlier chapters, you must be familiar with a number of terms, beginning with the term *analytics*.

Analytics

CONCEPT Descriptive and inferential methods that focus on discovering patterns to or generating hypotheses about data already collected.

INTERPRETATION The statistical methods in earlier chapters rely on first identifying and selecting relevant variables and then having some understanding of the relationship among variables. Those methods cannot uncover a pattern that has not been first specified, nor can they suggest variables that might be more significant to study than the ones already chosen. Those methods often represent a snapshot-in-time, analyzing data that has been collected at a much earlier time. Analytics methods remove these restrictions and allow new patterns to be uncovered and, in many cases, facilitate the analysis of data as the data is being collected (or very soon afterward).

Analytics methods can be divided into three categories: descriptive analytics, predictive analytics, and prescriptive analytics. Chapter 13 discusses descriptive analytics, and Chapter 14 covers predictive analytics.

Big Data

CONCEPT High-volume, high-velocity, and/or high-variety collections of data that require innovative forms of information processing for enhanced insight and decision making.

EXAMPLES A retailer might combine a customer's purchase history with the customer's credit card information, social media correspondence to the retailer, and demographic information collected by others. A professional sports team might track sales, movement, and other attendee interactions through a sports arena complex, from parking lot to stadium seat, during game day.

INTERPRETATION Big Data are collections of data that cannot be handled in traditional ways and that require advanced information processing techniques. You cannot open Big Data as you could open one of the data sets that are identified and used in this book. Because the term was originally coined by a consulting firm, the exact definition of Big Data can vary, but most agree that Big Data must show some combination of the following "3 V's":

- **Volume**: The collection of data is very large. For the retailer Big Data example, the combined customer data could be for millions of customers, each of whom could generate dozens of transactions a year, each of which could have multiple items.
- **Velocity**: The data is generated (and used) at a very fast rate. In the sports team Big Data example, every minute (or less), the status of an attendee could change, as that person arrives in a parking lot, passes through a turnstile, buys something from a concession or merchandise stand, and finds the proper seat.
- **Variety**: The way data are stored varies. For the retailer Big Data example, some of the data used, such as store transaction receipts, could be easily entered into a table or worksheet format, whereas other types of data, such as the contents of a customer's email to the retailer, could not be.

The capabilities of analytics methods to discover patterns to or generate hypotheses about data are well suited to Big Data, in which identifying relevant variables and or having new insights about relationships among variables would otherwise be hard to do. Therefore, Big Data and analytics are often mentioned together, even though analytics methods can also be used with smaller data sets.

Descriptive Analytics

CONCEPT Descriptive methods that summarize large collections of data or variables to present constantly updated status information about something.

EXAMPLES The real-time status display for all phases of a sports arena operation; a display that compares the current value of a variable to its range of values over time and a target.

INTERPRETATION Descriptive analytics summarize status or historical data about an organization. They extend the methods discussed in Chapters 2 and 3 by providing more than a fixed snapshot about one past moment in time. Individual methods can be combined into one *dashboard* that serves as a master summary of some business process or objective and that can be constantly updated. Many methods better summarize Big Data because these methods can visually summarize more than one variable at the same time, thereby providing clues about relationships among variables that might otherwise not be seen. Descriptive analytics methods typically incorporate **drill-down**, the capability to see ever-decreasing levels of summarization. (Chapter 13 discusses drill-down.)

Predictive Analytics

CONCEPT Methods that identify what is likely to occur in the near future and find relationships in data that might not be readily apparent using descriptive analytics.

EXAMPLES Predicting customer behavior in a retail context, detecting patterns of fraudulent financial transactions.

INTERPRETATION Predictive methods assign a value to a target based on a model, classify items in a collection to target categories or classes, group or cluster items into natural groupings (clusters), or find items that tend to co-occur and specify the rules that govern their co-occurrence.

In one famous example, the retailer Target sought to identify shoppers who were pregnant but who chose not to disclose that fact directly to Target. By looking at specific products such as skin lotions, nutritional supplements, and scent-free soaps, the retailer was able to create a predictive model of the purchase patterns for about 25 products to predict whether a customer was likely to be pregnant. In an early test of this model, the model was used to create mailings of coupons for pregnancy-related and baby products. One indignant father of a teenage girl who received such coupons came to a Target store asking why the coupons were mailed to *his* daughter, only to find out later from her that she was pregnant.

Data Mining

CONCEPT Predictive analytics method that looks for previously unknown patterns in Big Data collections.

EXAMPLES Identification of Target customers likely to be pregnant; an online consumer website's recommendation lists, such as the Netflix streaming service's recommendation engine.

INTERPRETATION The term *data mining* comes from analogy with natural resource mining. Unlike statistical methods discussed earlier that require selecting samples as a first step, data mining seeks to explore ("mining") the data resource that an organization has already accumulated in its own Big Data repository (the "mine"). Data mining can perform any one of the four types of predictive analysis.

Prescriptive Analytics

CONCEPT Methods that can suggest the best future decision making for specific case situations.

EXAMPLES Recommended daily optimal staffing levels for retailers that minimize labor costs but maximize sales; a system for a tour operator to set hotel pricing to maximize revenue while ensuring pricing consistency across all levels of accommodation offered.

INTERPRETATION Prescriptive methods seek to improve performance by optimizing decisions. Forerunners of these methods include the classical *optimization methods* used in management science and operations research. Prescriptive analytics leverage descriptive and predictive methods to create the basis for recommendation.

12.2 Software for Analytics

Current versions of Microsoft Excel include some descriptive analytics methods, but most analytics processing requires advanced statistical software or specialized analytics programs. Although these programs can cost a significant amount to own or lease, many come with either a free, time-limited trial version or a "public" version that is free for individuals to use with smaller sets of data. (Using analytics with Big Data requires specialized data storage software found in larger computer systems.) Chapter 13 uses Excel for its "Spreadsheet Solutions," but Chapter 14 uses JMP, the interactive data analysis program from the SAS Institute, Inc., to work out problems. Although a full discussion of JMP is beyond this book, "Software Solutions" in that chapter summarizes the JMP techniques used.

One-Minute Summary

Basic Vocabulary of Analytics

- Analytics: descriptive, predictive, and prescriptive
- Big Data
- Data mining

Test Yourself

Short Answers

1. Methods that focus on discovering patterns to or generating hypotheses about data already collected are called
   (a) Confidence intervals
   (b) Regression
   (c) Analysis of variance
   (d) Analytics
2. Big Data usually involves
   (a) High-volume data
   (b) Data that is generated at a fast rate
   (c) Data that is stored in a variety of ways
   (d) All of the above
3. Methods that summarize large collections of data and present constantly updated status information are called
   (a) Descriptive analytics
   (b) Predictive analytics
   (c) Prescriptive analytics
   (d) All of the above

4. Methods that identify what is likely to occur in the near future and find relationships in data that might not be readily apparent are called
   (a) Descriptive analytics
   (b) Predictive analytics
   (c) Prescriptive analytics
   (d) All of the above

5. Methods that can suggest the best future decision making for specific case situations are called
   (a) Descriptive analytics
   (b) Predictive analytics
   (c) Prescriptive analytics
   (d) All of the above

Answer True or False:

6. Analytics focuses on discovering patterns to or generating hypotheses about data already collected.
7. Big Data involves large amounts of data.
8. Big Data involves data that is collected infrequently over long periods of time.
9. Big Data involves data that is usually collected in a single format.
10. Descriptive analytics methods summarize large collections of data and present constantly updated status information.
11. Predictive analytics methods identify what is likely to occur in the near future and find relationships in data that might not be readily apparent.
12. Prescriptive analytics methods identify what is likely to occur in the near future and find relationships in data that might not be readily apparent.
13. Data mining usually involves data that has already been stored in a company's Big Data repository.
14. Most analytics methods require the use of specialized statistical software.
15. Currently, Microsoft Excel can be used for all descriptive analytics methods and predictive analytics methods.

Answers to Test Yourself Short Answers

1. d
2. d
3. a
4. b
5. c
6. True
7. True
8. False
9. False
10. True
11. True
12. False
13. True
14. True
15. False

References

1. Braun, Vivian. "Prescriptive versus Predictive: An IBMer's Guide to Advanced Data Analytics in Travel." Tnooz, **bit.ly/1yVJMZe**.
2. Duhigg, Charles. "How Companies Learn Your Secrets." *The New York Times Sunday Magazine*, 20 February 2012.
3. "Gartner IT Glossary." **www.gartner.com/it-glossary/big-data**.
4. Kimball, Aaron. "Understanding What Big Data Can Deliver." *Dr. Dobb's Journal*, November 2013.
5. Levin, Yuri. "Prescriptive analytics: the key to improving performance." SAS Knowledge Exchange, **bit.ly/1pa4H6L**.
6. "NASDAQ Wall Capabilities." **www.nasdaq.com/reference/wall_cap.stm**.

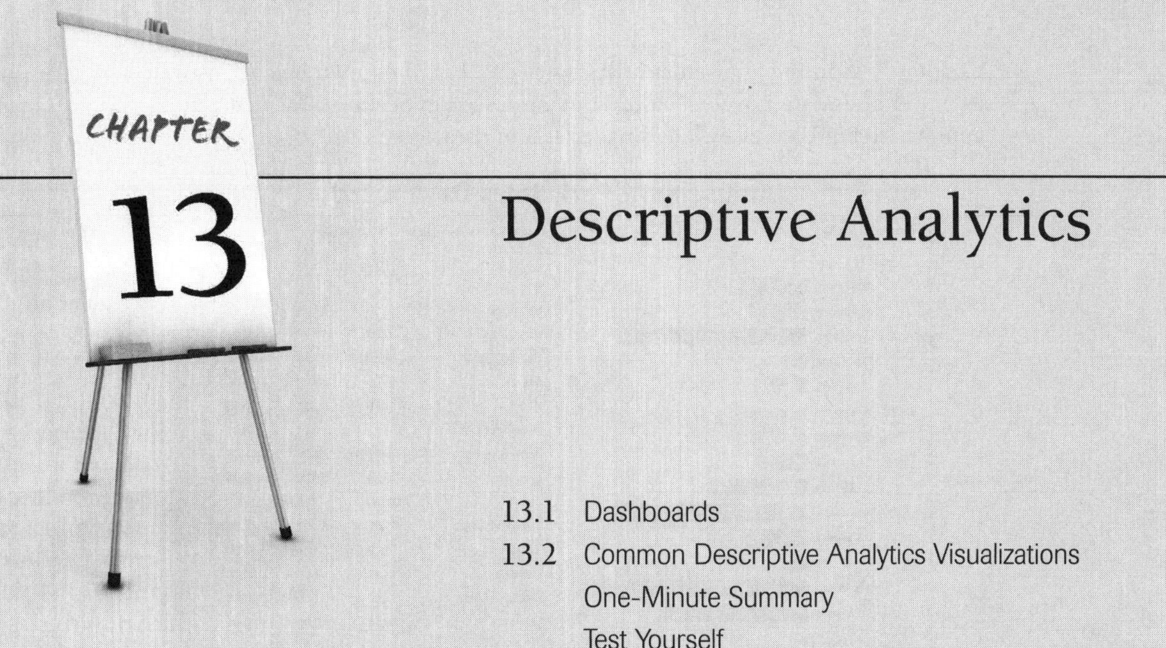

Descriptive Analytics

Descriptive analytics summarize large collections of data to present constantly updated status information or provide historical context. This chapter explores some of the common descriptive analytics methods, with an emphasis on techniques that can be used in Microsoft Excel.

13.1 Dashboards

CONCEPT Visual display that summarizes the most important variables needed for decision making or achieving a business objective.

EXAMPLES The real-time status display for all phases of a sports arena operation; the NASDAQ MarketSite Video Wall that summarizes trends and transactions at the NASDAQ securities exchange.

INTERPRETATION For several decades, people talked about developing executive information systems that would put information at the "fingertips" of decision makers. An analytics dashboard provides this information in a visual form that is intended to be easy to comprehend and review. Dashboards can contain the summary tables and charts discussed in

Chapter 2, as in the amusement park sales dashboard (shown below), as well as newer and more novel visualizations discussed in Section 13.2 that can summarize Big Data as well as smaller sets of data.

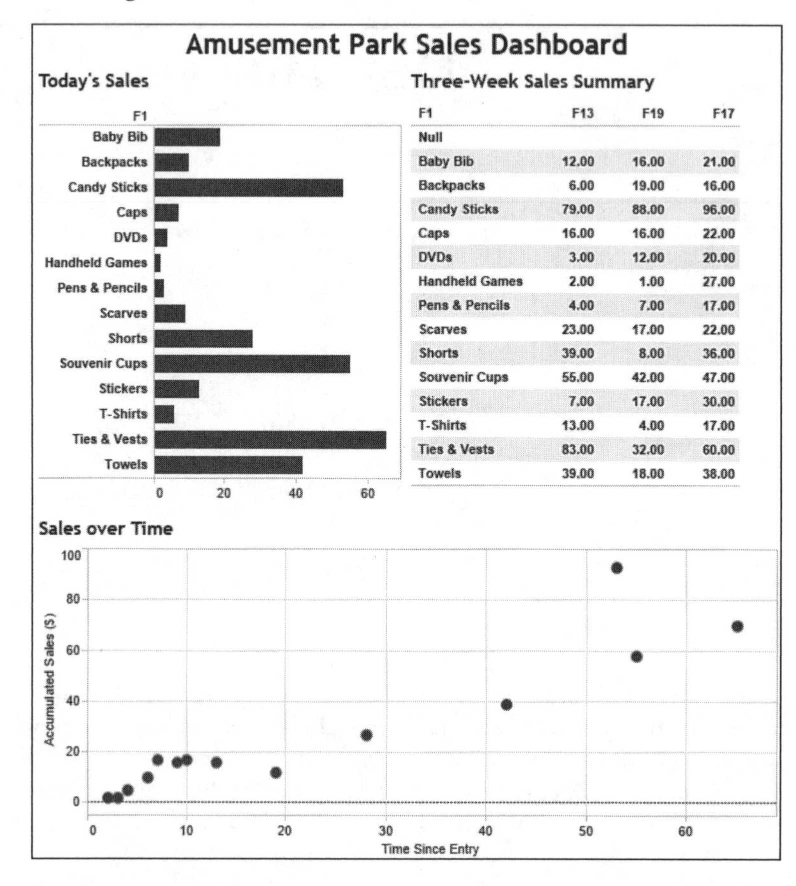

F1	F13	F19	F17
Null			
Baby Bib	12.00	16.00	21.00
Backpacks	6.00	19.00	16.00
Candy Sticks	79.00	88.00	96.00
Caps	16.00	16.00	22.00
DVDs	3.00	12.00	20.00
Handheld Games	2.00	1.00	27.00
Pens & Pencils	4.00	7.00	17.00
Scarves	23.00	17.00	22.00
Shorts	39.00	8.00	36.00
Souvenir Cups	55.00	42.00	47.00
Stickers	7.00	17.00	30.00
T-Shirts	13.00	4.00	17.00
Ties & Vests	83.00	32.00	60.00
Towels	39.00	18.00	38.00

Drill-Down

CONCEPT The revelation of the data that underlies a higher-level summary.

EXAMPLES Clicking an entry in an Excel PivotTable to reveal the worksheet rows that the entry summarizes; clicking a dashboard visual to see more detail about the summary information being displayed.

INTERPRETATION Drill-down is an important feature of dashboards and many other descriptive analytics methods. Drill-down helps manage the complexity or detail of the data being summarized, showing only the level of complexity and detail necessary for individual decision makers. (Different decision makers have different needs.) Drill-down also permits *data discovery*, the process by which decision makers can review data for patterns or exceptional or unusual values or outliers.

WORKED-OUT PROBLEM 1 You have been asked to further analyze a sample of mutual funds that have been classified by investment type (growth or value), risk (low, average, or high), and market cap (small, midcap, or large) using a PivotTable that summarizes the mean 10-year return percentages by type and risk (see table at left below).

You expand the table by drilling down to see the breakdown of growth and value funds by market cap (see table at right below). The original summary table shows that value funds with low or high risk have a higher mean 10-year return percentage than the growth funds with those risk levels, but the expanded table reveals a more complicated pattern. This table shows that growth funds with large market capitalizations are the poorest performers and significantly depress the mean for the growth fund category. (The blank cell additionally reveals that, in this sample, there are no value funds with high risk and a large market cap.)

Contingency Table of Fund Type, Market Cap, and Risk, showing the mean ten-year return percentage

Mean 10YrReturn%	RISK			
TYPE	Low	Average	High	Grand Total
⊞Growth	4.12	5.07	4.72	4.73
⊞Value	5.14	4.71	6.87	5.47
Grand Total	4.50	4.99	5.48	4.95

Contingency Table of Fund Type, Market Cap, and Risk, showing the mean ten-year return percentage out

Mean 10YrReturn%	RISK			
TYPE	Low	Average	High	Grand Total
⊟Growth	4.12	5.07	4.72	4.73
Large	3.69	3.65	1.26	3.48
Mid-Cap	5.62	6.04	5.77	5.92
Small	5.38	6.15	5.30	5.65
⊟Value	5.14	4.71	6.87	5.47
Large	4.52	4.13		4.34
Mid-Cap	6.62	6.27	5.52	6.01
Small	10.77	8.12	7.58	7.94
Grand Total	4.50	4.99	5.48	4.95

When you further drill down to analyze additional variables of the midcap funds with low risk, you discover another difference between the growth and value funds. For midcap value funds with low risk, the funds with the largest number of assets have the lowest expense ratios. This inverse relationship does not hold for growth funds in which the lowest expense ratios are associated with funds with modest asset sizes.

Fund Number	Market Cap	Type	Assets	Turnover Ratio	Beta	SD	Risk	10YrReturn%	Expense Ratio
RF241	Mid-Cap	Value	44.7	21	1.02	18.99	Low	5.43	1.27
RF239	Mid-Cap	Value	1452.0	43	0.93	17.47	Low	8.2	1.01
RF238	Mid-Cap	Value	196.9	68.05	0.76	15.86	Low	6.68	1.70
RF235	Mid-Cap	Value	1546.1	41	0.99	18.64	Low	8.9	1.15
RF231	Mid-Cap	Value	37.1	82	0.86	16.31	Low	3.88	1.76

Fund Number	Market Cap	Type	Assets	Turnover Ratio	Beta	SD	Risk	10YrReturn%	Expense Ratio
RF222	Mid-Cap	Growth	70.5	205	0.59	13.31	Low	2.07	1.81
RF221	Mid-Cap	Growth	150.6	150	0.82	16.16	Low	7.82	1.30
RF217	Mid-Cap	Growth	135.4	7	0.76	15.42	Low	0.53	1.31
RF216	Mid-Cap	Growth	9.1	246	0.95	18.4	Low	2.68	1.17
RF208	Mid-Cap	Growth	110.6	27.93	0.88	17.24	Low	4.88	0.99
RF207	Mid-Cap	Growth	3507.4	18	0.93	18.02	Low	9.76	1.14
RF203	Mid-Cap	Growth	174.0	12	0.92	17.52	Low	5.99	1.25
RF202	Mid-Cap	Growth	61.8	17.99	0.98	18.46	Low	6.18	1.46
RF200	Mid-Cap	Growth	287.6	16	0.95	18.59	Low	6.73	1.21
RF190	Mid-Cap	Growth	27.9	159	0.79	16.19	Low	8.06	2.00
RF188	Mid-Cap	Growth	319.5	7	0.96	18.1	Low	7.3	1.23
RF184	Mid-Cap	Growth	95.4	35	0.92	17.92	Low	5.38	1.04

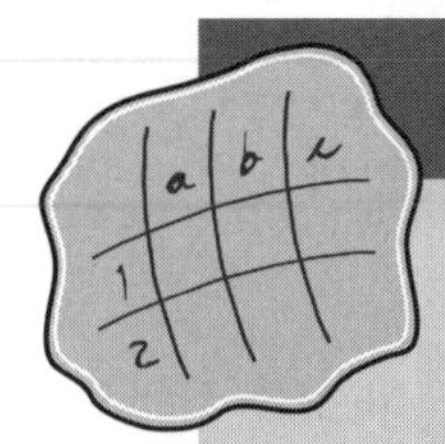

spreadsheet solution

Drill-Down

Chapter 13 Drilldown contains the drill-down steps of WORKED-OUT PROBLEM 1 as separate worksheets. Explore the set of worksheets to see the various levels of detail revealed by drilling down in the original PivotTable. Experiment by double-clicking the various PivotTable cells in the ExpandedPivotTable worksheet to produce an additional worksheet that reveals all the details about a particular group of mutual funds.

Best Practices

Use the Excel PivotTable feature to create tables into which you can drill down.

13.2 Common Descriptive Analytics Visualizations

Besides the tables and charts in Chapter 2, you can use some visually oriented descriptive analytics methods used alone or as part of a dashboard display. Most commonly encountered are sparklines, treemaps, and bullet graphs.

Sparklines

CONCEPT Chart that visualizes measurements taken over time as small, compact graphs designed to appear as part of a table or written passage.

EXAMPLE A sparklines chart that presents a four-year historical context to yearly changes in new car sales for four manufacturers (companies A, B, C, and D).

Company	Four-Year Sparklines
A	
B	
C	
D	

INTERPRETATION First described by Edward R. Tufte, sparklines provide a historical context for measurements taken over time, data known as *time-series* data. This context can override false impressions that can occur when single summary measures are displayed.

WORKED-OUT PROBLEM 2 You want to better understand the yearly percent changes in new car sales shown by the summary table (below).

Company	Yearly Change
A	Up 3.0%
B	Up 33.2%
C	Up 6.3%
D	Up 11.8%

You examine the four-year sparklines for the four manufacturers shown on page 268 in which the last year corresponds to the year to which the summary table refers. Those sparklines reveal, among other things, that company B's exceptional yearly change reflects a severe drop in sales in the previous year, company D's sales have greatly declined over the four years, and company C's sales have been least subject to fluctuations during the four-year period.

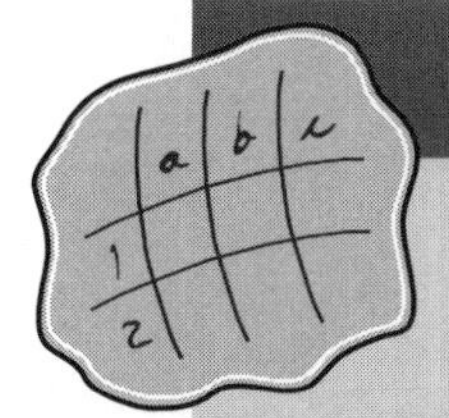

spreadsheet solution

Sparklines

Chapter 13 Sparklines contains the sparklines table shown on page 268 that displays the historical context of changes in new car sales used in WORKED-OUT PROBLEM 2. Experiment with this chart by changing the yearly sales data in columns B through E.

Best Practices

Use **Line Sparkline** on the **Insert tab** to create a sparklines table. After Excel creates the table, select **Design → Axis** and click **Same for All Sparklines** under Vertical Axis Minimum Value Options. Select **Design → Axis** a second time and click **Same for All Sparklines** under Vertical Axis Maximum Value Options.

Treemap

CONCEPT Chart that visualizes the comparison of two or more variables using the size and color of rectangles to represent values.

INTERPRETATION The varying size and color of the rectangles in a treemap can serve as a quick preliminary means of analysis. When used with more categorical variables, a treemap forms a multilevel hierarchy or tree that can uncover patterns among numerical variables inside each category.

Treemaps represent a tradeoff between precision and quick impressions. In decision-making situations that concern managing the status of some activity, the latter can be more useful *if* the variables that the treemap visualizes are thoughtfully chosen.

WORKED-OUT PROBLEM 3 You have been asked to form some preliminary observations from a treemap that compares the assets and 10-year rates of return of growth and value funds selected for a sample. The treemap (below) quickly shows that growth funds (left side of treemap) and value funds (right side of treemap) tend to show similar relationships between asset size and 10-year rates of return. Specifically, you can observe that, generally, the smallest funds have the worst (lightest-color) returns, while "middle-sized" funds seem to have the best (darkest-color) returns.

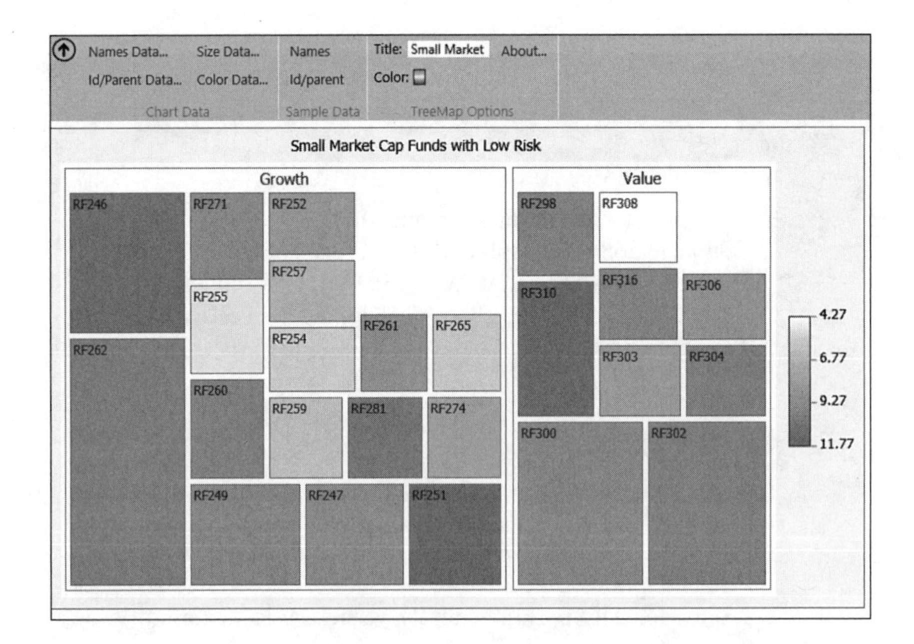

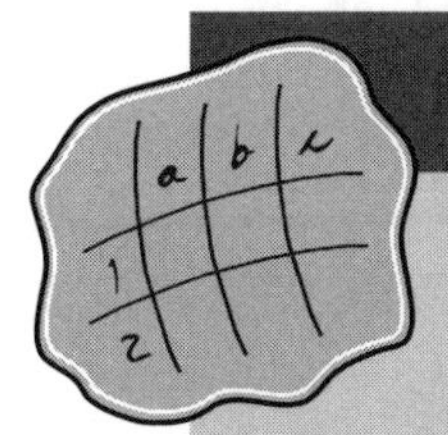

spreadsheet solution

Treemaps

Chapter 13 Treemap contains the treemap shown on page 271 and used in WORKED-OUT PROBLEM 3. Exploring or experimenting with the treemap requires the Treemap app. If your copy of Excel does not have this app installed, the treemap will appear faded and an informational message explaining how to download and use the app will appear.

Best Practices

To download and install the Treemap app for Excel, click the **Store icon** on **the Insert tab** to connect to the Microsoft Office apps store (requires Internet access). Follow the instructions that are included in the app's home page to complete the installation.

Bullet Graph

CONCEPT Bar chart that combines the visualization of a single numerical variable (the bar) with a categorical variable (shadings behind the bar) that helps to classify the numerical value being shown.

INTERPRETATION Stephen Few devised bullet graphs as a compact visualization for dashboards. Bullet graphs enhance bar charts by providing a context for the numerical value being graphed. This context takes the form of ranges of values that are described by categories, such as Poor, Acceptable, and Excellent. Bullet graphs can also include additional complementary measures, such as target values that appear as lines or symbols on the same axis as the bars. For example, the bullet graph on the next page displays the weekly sales of items from the souvenir stands at an amusement park. In this visualization, each type of item has its own sales target (in units, vertical line) and ranges of what constitutes poor (dark shade), acceptable (light shade), or excellent sales (white background).

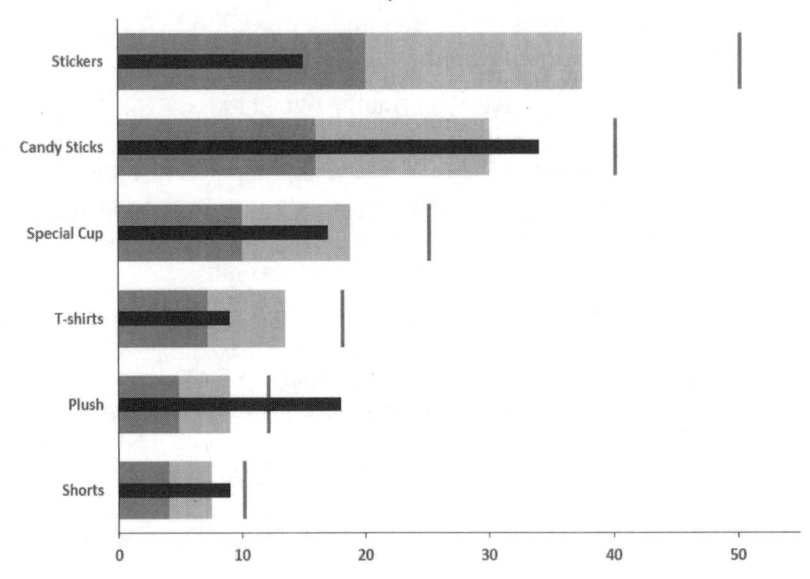

WORKED-OUT PROBLEM 4 You have been asked to characterize the weekly sales for items sold in the souvenir stands at an amusement park. Using the previous bullet graph, you show that sales of plush toys have exceeded its target, candy sticks and shorts are selling well, and sales of stickers are doing poorly.

Bullet Graphs

Chapter 13 Bullet Graph contains the bullet graph shown on this page that summarizes weekly unit sales of souvenir stands, used in WORKED-OUT PROBLEM 4. Experiment with this chart by changing the target and actual values in columns B and C of the BulletData worksheet.

Best Practices

Because bullet graphs are not a chart type included in Excel, use a custom combination chart in which the actual values (the black bars) are plotted on a secondary axis with bars that are narrower, with a gap width between bars of 500%.

One-Minute Summary

Descriptive analytics foundation

- Dashboard
- Drill-down

What visualization to use?

- To provide a historical context for a statistic, use sparklines.
- To provide a preliminary comparison among variables divided into categorical groups, use a treemap.
- To present a set of related numerical values and provide a reference by which those values can be characterized and, optionally, compared to a reference value, use a bullet graph.

Test Yourself

1. A visual display that summarizes the most important variables needed for decision making or achieving a business objective is called a
   (a) Dashboard
   (b) Sparkline
   (c) Treemap
   (d) Bullet graph
2. A display that visualizes measurements taken over time as small, compact graphs designed to appear as part of a table or written passage is called a
   (a) Dashboard
   (b) Sparkline
   (c) Treemap
   (d) Bullet graph
3. A chart that visualizes the comparison of two or more variables using the size and color of rectangles to represent values is called a
   (a) Dashboard
   (b) Sparkline
   (c) Treemap
   (d) Bullet graph

4. A bar chart that combines the visualization of single numerical variable (the bar) with a categorical variable (shadings behind the bar) is called a
   (a) Dashboard
   (b) Sparkline
   (c) Treemap
   (d) Bullet graph

5. Drill-down
   (a) Can show various levels of summarization
   (b) Helps manage the complexity of a set of data
   (c) Is an important feature of dashboards
   (d) All of the above

Answer True or False:

6. Dashboards summarize a set of variables needed for decision making.
7. Drilling down enables you to break down a variable according to the values of another variable to see patterns that might not be apparent in the original variable.
8. Drilling down always reveals more details about data being summarized.
9. Sparklines display measurements for a single point of time.
10. Sparklines display measurements over a period of time.
11. Sparklines can display measurements for several groups on a single graph.
12. Treemaps visualize the comparison of two or more variables using the size and color of rectangles to represent values.
13. The color on a treemap represents the same variable as the size of a category.
14. A bullet graph compares variables that have been divided into categorical groups.
15. The color on a bullet graph represents the same variable as the length of the bars.

Answers to Test Yourself

1. a
2. b
3. c
4. d
5. d
6. True
7. True
8. True
9. False
10. True
11. True
12. True
13. False
14. False
15. False

References

1. Few, Stephen. *Information Dashboard Design*, 2nd ed. Burlingame, CA: Analytics Press, 2013.
2. *Microsoft Excel* 2013. Readmond, WA: Microsoft Corporation, 2012.
3. "NASDAQ Wall Capabilities." **www.nasdaq.com/reference/wall_cap.stm**.
4. Perceptual Edge. *Bullet Graph Design Specification*. **bit.ly/1pal7f9**.
5. Tufte, Edward. *Beautiful Evidence*. Cheshire, CT: Graphics Press, 2006.

CHAPTER 14

Predictive Analytics

Predictive analytics methods identify what is likely to occur in the near future and find relationships in data. In this chapter, you learn the types of problems to which predictive analytics can be best applied and learn the specifics of several predictive analytics methods.

14.1 Analysis with Predictive Analytics

Predictive analytics methods can be applied to four types of analysis: predicting, classifying, clustering, and associating data.

Predicting

CONCEPT Analyses that assign a value to a target based on a model.

EXAMPLE Predicting the chance that someone will default on a mortgage based on various characteristics of the individual.

Classifying

CONCEPT Analyses that assign items in a collection of items to target categories or classes.

EXAMPLE Assigning credit card holders to the category of likely to upgrade to a premium card, based on the number of additional cards they have and the amount they charge on the card per month.

Clustering

CONCEPT Analyses that discover natural groupings, called *clusters*, in a set of data.

EXAMPLE Grouping different types of foods into clusters, based on the perceived characteristics of those food types.

Associating

CONCEPT Analyses that find sets of items that tend to co-occur and that specify the rules that govern their co-occurrence.

EXAMPLE Using the perceived characteristics of each type, mapping different types of foods in two or more dimensions.

Specific methods of predictive analytics can employ one or more types of analysis. The following table shows which analysis types apply to the specific methods discussed in the rest of this chapter.

	Predicting	Classifying	Clustering	Associating
Tree analysis (classification and regression trees)	•	•		
Cluster analysis			•	
Multidimensional scaling		•		•

14.2 Classification and Regression Trees

CONCEPT Classification and regression tree analysis splits data into groups based on the values of independent or explanatory (X) variables.

INTERPRETATION Splitting data into groups one or more times can be visualized as a tree diagram that branches into forks at each level or *node*. Tree analysis seeks to determine which values of a specific independent variable are useful in predicting the dependent (Y) variable. When using a tree analysis method, you do not first determine which variables to include in the model—the method itself determines the variables to be used. To construct a tree, classification and regression tree methods choose the one independent variable that provides the best split of the data at each node in the tree, starting with the root. These methods must also employ rules for deciding when a branch cannot be split anymore.

Classification and regression tree methods are not affected by the distribution of the variables that make up the data. Typically, trees are developed through several levels of nodes until either no further gain in the fit occurs or the splitting has continued as far as possible. After splitting, some methods *prune* the tree, deleting splits that do not enhance the final analysis.

Training, Validation, and Test Sets

As with all other methods used for prediction, classification and regression tree methods work best when you employ training and validation sets to refine the model. These two sets are formed by splitting the original set of data under analysis. The training set helps a tree method establish the tree, and the validation set helps refine the tree by making sure the tree does not too precisely model only the training data. If the original set of data is large enough, the set is often divided into three sets: training, validation, and test. In this case, the test set serves as the final evaluation of the model that the tree analysis creates. When the original set of data is small, all the data can be used as the training set (and no validation set would be used).

Classification Trees

CONCEPT The tree analysis in which the dependent (Y) variable to be predicted is categorical.

WORKED-OUT PROBLEM 1 You want to develop a classification tree model to predict the probability that a customer will be satisfied, based on the delivery time of a room service breakfast. You collect the results for 30 deliveries on a particular day and record whether the customer was satisfied, the difference between the actual and requested delivery times (a negative time means that the breakfast was delivered before the requested time), and whether the customer previously stayed at the hotel.

Because the sample is so small, you decide to use all the data as the training set. The results for these data using classification tree software are shown on the next page. The r^2 for the classification tree model is 0.573. This means that 57.3% of the variation in satisfaction can be explained by variation in the delivery time difference and whether the customer previously stayed at the hotel.

Satisfaction

Satisfaction	Del. Time Difference	Previous	Satisfaction	Del. Time Difference	Previous
No	6.1	Yes	Yes	3.6	Yes
Yes	4.5	No	No	6.0	No
Yes	0.8	No	Yes	4.4	Yes
Yes	1.3	Yes	Yes	0.9	No
Yes	3.6	Yes	Yes	1.2	Yes
Yes	2.7	Yes	No	3.8	No
No	5.9	No	Yes	3.5	Yes
Yes	4.5	Yes	No	4.0	Yes
No	4.8	Yes	Yes	4.3	Yes
Yes	2.1	No	No	4.9	Yes
Yes	4.1	Yes	Yes	2.3	No
No	5.6	No	Yes	3.8	Yes
Yes	3.8	Yes	No	5.9	Yes
Yes	2.3	No	Yes	3.7	Yes
No	3.2	No	Yes	2.5	No

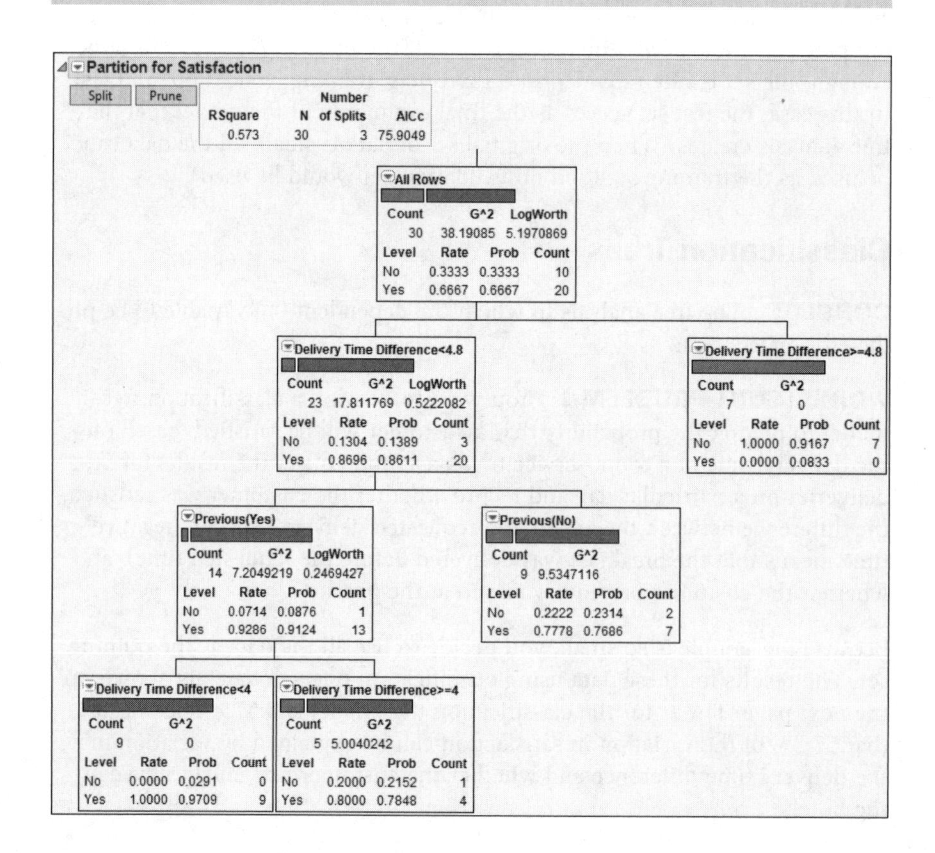

The first split is for customers who experienced a delivery time difference of less than 4.8 minutes. Of the 23 customers who had a delivery time difference of less than 4.8 minutes, 20 were satisfied and 3 were not. All seven customers who had a delivery time difference greater than 4.8 minutes were not satisfied. This model suggests that the most important determinant of satisfaction was having a delivery time difference of below 4.8 minutes.

The second split was among those customers who had a delivery time difference of less than 4.8 minutes. Fourteen of these customers had previously stayed at the hotel. Of these 14 customers, 13 were satisfied and 1 was not. Of the nine customers who had not stayed at the hotel, seven were satisfied and two were not.

The third split was among the 14 customers who previously stayed at the hotel. All nine of these customers who had a delivery time difference of less than 4 minutes were satisfied. Of the five customers who had a delivery time difference of more than 4 minutes, four were satisfied and one was not. This split suggests that customers who had previously stayed at the hotel were more inclined to be satisfied, especially if the delivery time difference was less than 4 minutes.

Regression Trees

CONCEPT The tree analysis in which the dependent (Y) variable to be predicted is numerical.

WORKED-OUT PROBLEM 2 You want to develop a regression tree model to predict the effect of radio and newspaper advertising on the sales of a new product. You allocate to each city a specific expenditure level for radio advertising and for newspaper advertising. During a one-month test period, you collect data from a sample of 22 cities with approximately equal populations. The new product sales (in $thousands) and radio and newspaper advertising expenses (both in $thousands) collected during the test month are as follows:

Advertise

Sales	Radio	Newspaper	Sales	Radio	Newspaper
973	0	40	1577	45	45
1119	0	40	1044	50	0
875	25	25	914	50	0
625	25	25	1329	55	25
910	30	30	1330	55	25
971	30	30	1405	60	30
931	35	35	1436	60	30
1177	35	35	1521	65	35
882	40	25	1741	65	35
982	40	25	1866	70	40
1628	45	45	1717	70	40

Because the sample is so small, you decide to use all the data as the training set. Results for these data using regression tree software are:

Partition for Sales

Split | Prune

RSquare	RMSE	N	Number of Splits	AICc
0.799	151.36448	22	2	293.653

All Rows

		LogWorth	Difference
Count	22		
Mean	1225.1364	6.4175871	514.5
Std Dev	345.5701		

Radio<45

Count	10
Mean	944.5
Std Dev	148.92821

Radio>=45

		LogWorth	Difference
Count	12		
Mean	1459	2.6365319	432
Std Dev	280.3086		

Newspaper<35

Count	6
Mean	1243
Std Dev	212.74398

Newspaper>=35

Count	6
Mean	1675
Std Dev	125.004

The r^2 of the regression tree model is 0.799. This means that 79.9% of the variation in sales can be explained by variation in radio and newspaper advertising.

The first split occurs at radio advertising expenditures of $45,000. With radio advertising expenditures of less than $45,000, the mean sales are $944,500. With radio advertising expenditures of at least $45,000, the mean sales are $1,459,000. This means that spending at least $45,000 on radio advertising had the largest effect on sales.

The second split occurs at newspaper advertising expenditures of $35,000. For cities in which the radio advertisement expenditures were at least $45,000, the mean sales were $1,243,000 if newspaper advertising expenditures were less than $35,000, or is $1,675,000 if newspaper advertising expenditures were at least $35,000. This means that if radio advertising expenditures amount to at least $45,000, newspaper advertising expenditures of at least $35,000 would affect sales, too.

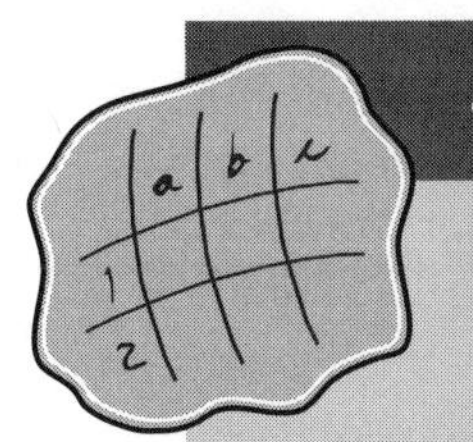

software solution

Classification and Regression Trees

The classification tree shown on page 280 and used in WORKED-OUT PROBLEM 1, and the regression tree shown on page 282 and used in WORKED-OUT PROBLEM 2 were created using the Partition procedure of the JMP program, available from the SAS Institute. The use of JMP is beyond the scope of this book; a trial version of JMP is available at **www.jmp.com**.

14.3 Cluster Analysis

CONCEPT Predictive analytics method that classifies data into a sequence of groupings so that objects of each group are more alike among themselves than they are to objects found in other groups.

INTERPRETATION In performing a cluster analysis, the results can be affected by the following features of clustering chosen:

- Hierarchical versus nonhierarchical clustering
- The measurement of the distance between objects
- The measurement of the distance between clusters

In hierarchical clustering, the analysis starts with each object in its own cluster. Then the two objects that are determined to be the closest to each other are merged into a single cluster. The merging of the two closest objects repeats until there remains only one cluster that includes all objects. In k-means clustering, which is a type of non-hierarchical clustering, the number of clusters (k) is set at the start of the process. Objects are then assigned to clusters in an iterative process that seeks to make the means of the k clusters as different as possible. During the iterative process, objects can be reassigned to a different cluster, unlike hierarchical clustering, in which clusters, once formed, are never changed later in the process.

The most common measure of distance between objects used in cluster analysis is Euclidean distance. This measure is based on the squared differences between objects taken over all the dimensions on which the objects are measured. The city block distance, based on the absolute value of the difference between objects taken over all the dimensions on which the objects are measured, is another measure of distance that is sometimes used.

Among the measures of distance between clusters are complete linkage, single linkage, average linkage, and Ward's minimum variance method. Complete linkage bases the distance between clusters on the maximum distance between objects in one cluster and in another cluster. Single linkage bases the distance between clusters on the minimum distance between objects in one cluster and in another cluster. Average linkage bases the distance between clusters on the mean distance between objects in one cluster and in another cluster. Ward's minimum variance method bases the distance between clusters on the sum of squares over all variables between objects in one cluster and in another cluster.

WORKED-OUT PROBLEM 3 Suppose you want to perform a cluster analysis to examine the similarities and dissimilarities among various sports. You conduct a survey and collect data about the perceptions people have about four attributes of nine sports (basketball, skiing, baseball, Ping-Pong, hockey, track and field, bowling, tennis, and U.S. football). You use the following 7-point rating scales for the four attributes:

- Movement speed: 1 = fast paced to 7 = slow paced
- Rules: 1 = complicated rules to 7 = simple rules
- Team orientation: 1 = team sport to 7 = individual sport
- Amount of contact: 1 = noncontact to 7 = contact

Consider the mean score of each sport on each rating scale:

Sports

Sport	Movement Speed	Rules	Team Oriented	Amount of Contact
Basketball	1.84	2.58	1.56	4.89
Skiing	3.98	5.17	5.96	3.01
Baseball	3.76	2.47	1.86	4.57
Ping-Pong	4.98	5.83	5.93	1.32
Hockey	1.71	2.22	1.82	5.96
Track and Field	5.83	4.88	5.47	2.11
Bowling	6.07	5.16	5.78	1.49
Tennis	4.67	4.26	5.47	2.16
Football (American)	2.92	2.26	1.44	6.47

Using cluster analysis software, results for these data are as follows:

Hierarchical Clustering

Method = Complete

Dendrogram

Basketball
Hockey
U.S. Football
Baseball
Skiing
Tennis
Ping-Pong
Track and Field
Bowling

Clustering History

Number of Clusters	Distance	Leader	Joiner
8	0.424470327	Track and Field	Bowling
7	0.616270380	Basketball	Hockey
6	0.901181166	Skiing	Tennis
5	0.955196619	Ping-Pong	Track and Field
4	1.078857527	Basketball	U.S. Football
3	1.483152744	Basketball	Baseball
2	1.530325339	Skiing	Ping-Pong
1	4.488847489	Basketball	Skiing

From examining either the tree diagram (which is also called a dendrogram) from left to right or the clustering history, observe that the first two sports that cluster together are track and field and bowling, followed by basketball and hockey. Then skiing and tennis join together, followed by Ping-Pong merging with both track and field and bowling. This process continues until all the sports are merged into one cluster.

When three clusters remain, the sports in the three clusters are {basketball, hockey, U.S. football, and baseball}, {bowling, Ping-Pong, and track and field}, and {tennis and skiing}. The first cluster of {basketball, hockey, U.S. football, and baseball} appears to represent team sports. The second cluster of {bowling, Ping-Pong, and track and field} consists of slow-moving, individually contested sports. The third cluster {tennis and skiing} represents fast-moving, individually contested sports.

software solution

Cluster Analysis

The cluster analysis shown on this page and used in WORKED-OUT PROBLEM 3 was created using the Cluster procedure of the JMP program, available from the SAS Institute. The use of JMP is beyond the scope of this book; a trial version of JMP is available at **www.jmp.com**.

14.4 Multidimensional Scaling

CONCEPT Predictive analytics method that visualizes objects in a map of two or more dimensions.

INTERPRETATION The goal of multidimensional scaling is to discover which objects are close to other objects on the map and what features characterize the dimensions of the map.

Among the issues that need to be resolved when doing a multidimensional scaling analysis are:

- The measurement of the distance between objects
- The scaling of the distance between objects
- The measure of the goodness of fit of the multidimensional scaling results in a varying number of dimensions

As with cluster analysis, the most common measure of distance between objects is Euclidean distance. This measure is based on the squared differences between objects taken over all the dimensions on which the objects are measured. The city block distance, based on the absolute value of the difference between objects taken over all the dimensions on which the objects are measured, is another measure of distance that is sometimes used.

The scaling between objects can be one of two types. With metric scaling, the distance between objects is assumed to be measured where there is a true zero point, and each distance is a ratio of another distance. With nonmetric scaling, the distance between objects is assumed to be based only on the ranks of the distances.

The most common goodness of fit statistic used in multidimensional scaling is the stress statistic. This statistic measures how well the results fit the actual distances between the objects. Although the smaller the stress statistic, the better the fit, no fixed rule constitutes an acceptable value for the stress statistic. As a general rule, the stress statistic decreases as the number of dimensions increases, so using many dimensions can cause the stress statistic to approach 0 (a perfect fit), but at the cost of creating a map that could be as complex as the original data itself. Usually, a good rule is to increase dimensions as long as the stress statistic decreases substantially. In many cases, the decrease in the stress statistic begins to fall off after the second or third dimension is considered, which is helpful because interpreting the significance of more than three dimensions can be challenging.

WORKED-OUT PROBLEM 4 Return to the survey that was conducted to examine the similarities and dissimilarities among various sports (see page 284). The results of the stress statistic for these data are as follows:

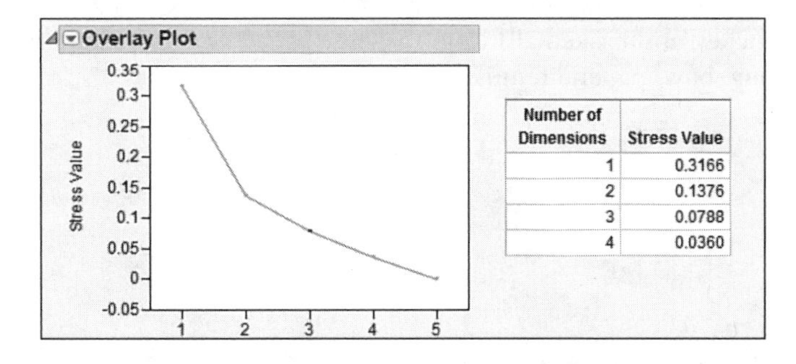

Number of Dimensions	Stress Value
1	0.3166
2	0.1376
3	0.0788
4	0.0360

The results reveal a stress statistic of 0.3166 in one dimension, 0.1376 in two dimensions, and 0.0788 in three dimensions. Because there is a large difference in the stress statistic between one and two dimensions, but only a small difference in the stress statistic between two and three dimensions, you choose to begin by interpreting the two-dimensional results shown below.

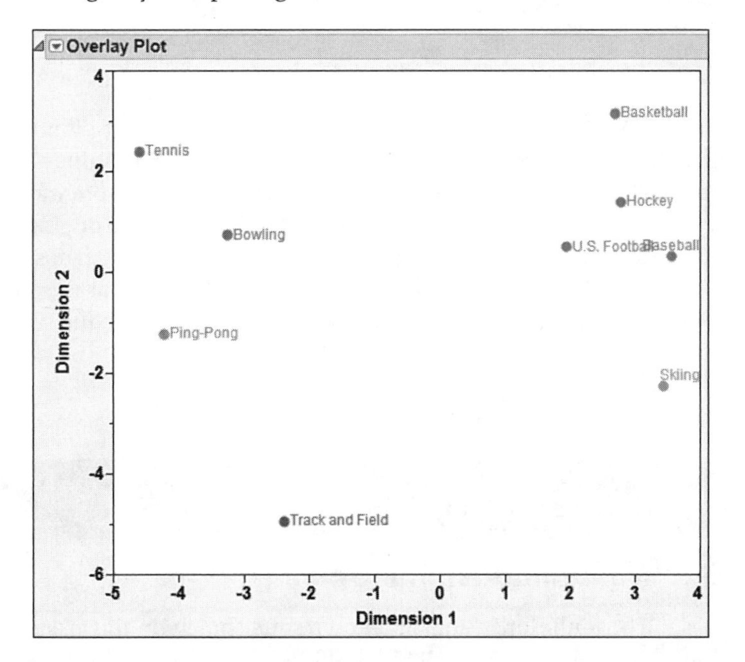

To interpret a two-dimensional map, you look for points that appear close to each other, as well as points that appear distant from each other. Although not the case with this figure, you might need to rotate the map to better interpret the dimensions. From this figure, observe that U.S. football, hockey, and basketball are close to each other. Tennis, Ping-Pong, and bowling are somewhat close to each other, and track and field and skiing are separate from the others.

To best interpret the dimensions separating the sports, observe that if you rotate the map clockwise 45 degrees, one axis appears to separate the team sports {U.S. football, hockey, basketball, and baseball} from the nonteam sports. The other axis appears to separate the fast-paced contact sports {U.S. football, hockey, and basketball} from the slow-paced noncontact sports, such as Ping-Pong, bowling, and tennis.

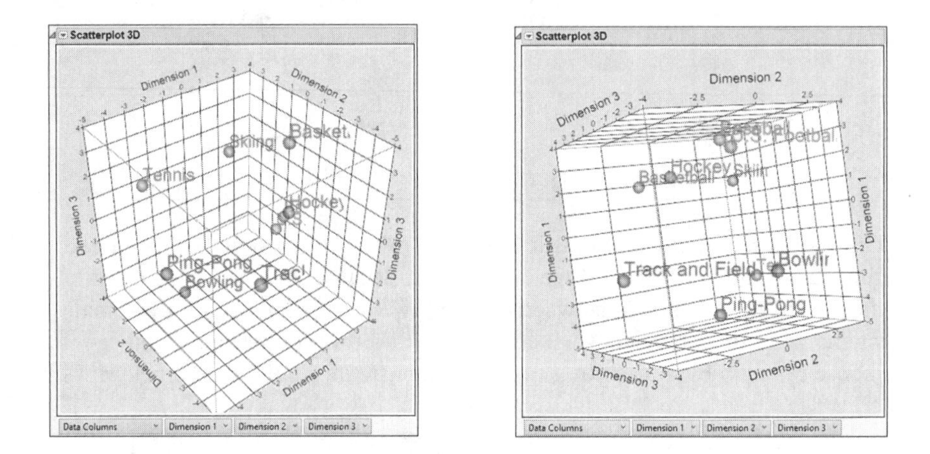

After interpreting the two-dimensional map, you can check to see if interpreting a three-dimensional map yields a better result. Interpreting a three-dimensional map is inherently harder because there are many more ways to examine and rotate the cubelike map. This last figure shows the original and rotated three-dimensional map. The rotated map seems to show team sports gathering near the "ceiling," with individual sports gathering near the "floor." Because this does not enhance the interpretation, you would use the simpler two-dimensional map for your final analysis.

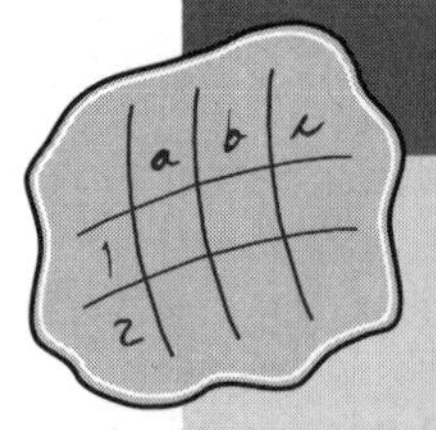

software solution

Multidimensional Scaling

The multidimensional scaling results shown on this page and used in WORKED-OUT PROBLEM 4 were created using the multidimensional scaling add-in for the JMP program, available from the SAS Institute. The use of JMP is beyond the scope of this book; a trial version of JMP is available at **www.jmp.com**.

One-Minute Summary

Predictive analytics

- Use a classification tree when the dependent (Y) variable to be predicted is categorical.
- Use a regression tree when the dependent (Y) variable to be predicted is categorical.
- Use cluster analysis to classify data into a sequence of groupings so that objects of each group are more alike among themselves than they are to objects found in other groups.
- Multidimensional scaling (MDS) visualizes objects in a map of two or more dimensions.

Test Yourself

1. A financial services company wants to be able to predict whether a cardholder would be willing to upgrade to a premium credit card. Among the independent variables to be considered are the monthly amount charged on the current credit card and the number of credit cards owned. Which of the following predictive analytics methods should you use to analyze these data?
   (a) Classification trees
   (b) Regression trees
   (c) Cluster analysis
   (d) Multidimensional scaling
2. A winery wants to study the factors involved in the quality rating of a wine. Data have been collected on various characteristics that can influence the quality rating of the wine. Which of the following predictive analytics methods should you use to analyze these data?
   (a) Classification trees
   (b) Regression trees
   (c) Cluster analysis
   (d) Multidimensional scaling
3. A historical researcher wants to better understand the demographic factors that might have increased the likelihood of surviving the sinking of the RMS Titanic in 1912. For each passenger on the ship, age, gender,

class of service booked, and whether the passenger was a survivor are collected. Which of the following analytics methods should you use to develop a model that would predict whether a passenger survived?

(a) Classification trees

(b) Regression trees

(c) Cluster analysis

(d) Multidimensional scaling

4. A basketball analytics specialist wanted to develop a model to predict the number of wins achieved by an NBA team. Among the independent variables used were field goal percentage, opponents' field goal percentage, three point field goal percentage, opponents' three point field goal percentage, rebounds, and turnovers. Which of the following predictive analytics methods should you use to analyze these data?

   (a) Classification trees

   (b) Regression trees

   (c) Cluster analysis

   (d) Multidimensional scaling

5. The owner of a restaurant serving continental food wants to study the perception that customers have of various entrée choices. Among the entrées available on the menu are steak, lamb chops, shrimp, lobster, scallops, halibut, trout, chicken, duck, and turkey. Each entrée was rated according to numerous characteristics, including spiciness, taste, healthfulness, and calories. The owner wants to determine which entrées are more similar to which other entrées and what characteristics seem to be separating the entrées from each other. Which of the following predictive analytics methods should you use to analyze these data?

   (a) Classification trees

   (b) Regression trees

   (c) Cluster analysis

   (d) Multidimensional scaling

6. The owner of a restaurant serving continental food wants to study the perception that customers have of various entrée choices. Among the entrees available on the menu are steak, lamb chops, shrimp, lobster, scallops, halibut, trout, chicken, duck, and turkey. Each entrée was rated according to numerous characteristics, including spiciness, taste, healthfulness, and calories. The owner wants to develop a map that shows the relative position of each entrée and be able to interpret the dimensions separating the entrées. Which of the following predictive analytics methods should you use to analyze these data?

   (a) Classification trees

   (b) Regression trees

   (c) Cluster analysis

   (d) Multidimensional scaling

7. A baseball analytics specialist wants to study the similarities and differences among the teams in a recent season. Numerous offensive and pitching statistics were obtained, including runs scored per game, home runs, batting average, runs allowed, earned run average, saves, and opponents' batting average. Which of the following predictive analytics methods should you use to analyze these data?
   (a) Classification trees
   (b) Regression trees
   (c) Cluster analysis
   (d) Multidimensional scaling

8. A baseball analytics specialist wants to study the similarities and differences among the teams in a recent season. Numerous offensive and pitching statistics were obtained, including runs scored per game, home runs, batting average, runs allowed, earned run average, saves, and opponents' batting average. The baseball analytics specialist wants to develop a map that shows the relative position of each team and be able to interpret the dimensions separating the teams. Which of the following predictive analytics methods should you use to analyze these data?
   (a) Classification trees
   (b) Regression trees
   (c) Cluster analysis
   (d) Multidimensional scaling

Answer True or False:

9. If the dependent (Y) variable is categorical, you can use a regression tree.

10. If the dependent (Y) variable is numerical, you can use a regression tree.

11. When using a classification or regression tree, the variables included in the model are determined in advance.

12. Once a tree is split, it is not possible to prune it back.

13. When you have a large enough set of data, you should divide the data into a training set, a validation set, and a test set.

14. In cluster analysis, objects in a group are more like other objects in their group than they are to objects in another group.

15. In hierarchical clustering, once two objects are assigned to a specific cluster, they cannot be subsequently separated into other clusters.

16. Euclidean distance used in cluster analysis and multidimensional scaling is based on the squared difference between objects taken over all the dimensions on which the objects are measured.

17. Complete linkage cluster analysis bases the distance between clusters on the maximum distance between objects in one cluster and in another cluster.
18. Single linkage cluster analysis bases the distance between clusters on the maximum distance between objects in one cluster and in another cluster.
19. Multidimensional scaling visualizes objects on a map of two or more dimensions.
20. As the number of dimensions decreases, the stress statistic in multidimensional scaling increases.

Answers to Test Yourself

1. a
2. b
3. a
4. b
5. c
6. d
7. c
8. d
9. False
10. True
11. False
12. False
13. True
14. True
15. True
16. True
17. True
18. False
19. True
20. True

References

1. Breiman, L., J. Friedman, C. J. Stone, and R. A. Olshen. *Classification and Regression Trees*. London: Chapman and Hall, 1984.
2. Cox, T. F., and M. A. Cox. *Multidimensional Scaling*, 2nd ed. Boca Raton, FL: CRC Press, 2010.
3. Everitt, B. S., S. Landau, and M. Leese. *Cluster Analysis*, 5th ed. New York: John Wiley, 2011.
4. *JMP Version* 11. Cary, NC: SAS Institute. 2013.
5. Lindoff, G., and M. Berry. *Data Mining Techniques: For Marketing, Sales, and Customer Relationship Management*. Hoboken, NJ: Wiley Publishing, Inc., 2011.
6. Loh, W. Y. "Fifty Years of Classification and Regression Trees." *International Statistical Review*, 2013.

APPENDIX A

Microsoft Excel Operation and Configuration

Use this appendix to learn how to operate and configure your copy of Microsoft Excel for use with this book. This appendix also reviews the conventions used in the book to describe various spreadsheet user operations.

Spreadsheet Operation Conventions

The spreadsheet operation instructions in this book use a standard vocabulary to describe keystroke and mouse (pointer) operations. Keys are always named by their legends. For example, the instruction "press Enter" means to press the key with the legend **Enter**.

For mouse operation, this book uses **click** and **select** and less frequently, **check**, **right-click**, and **double-click**. Click means to move the pointer over an object and press the primary mouse button. Select means to either find and highlight a named choice from a pull-down list or fill in an option (also known as radio) button associated with that choice. Check means to fill in the check box of a choice by clicking in its empty check box. Right-click means to press the secondary mouse button (or to hold down the Control key and press the mouse button, if using a one-button mouse). Double-click means to press the primary mouse button rapidly twice to select an object directly.

Spreadsheet Technical Configurations

The instructions in this book for using Microsoft Excel assume no special technical settings. If you plan to use any of the Analysis ToolPak Tips of Appendix E, you need to make sure that the Analysis ToolPak add-in has been installed in your copy of Microsoft Excel. (The Analysis ToolPak is *not included* with and is *not available* for Mac Excel 2008.)

To check for the presence of the Solver (or Analysis ToolPak) add-in, if you use Microsoft Excel with Microsoft Windows, follow these steps:

1. Select **File —> Options**. (In Excel 2007, click the **Office Button** and then click **Excel Options**.)

In the Excel Options dialog box:

2. Click Add-Ins in the left pane and look for the entry **Analysis ToolPak** in the right pane, under **Active Application Add-ins**.
3. If the entry appears, click **OK**.
4. If the entry does not appear in the Active Application Add-Ins list, select **Excel Add-Ins** from the **Manage** drop-down list and then click **Go**.
5. In the Add-Ins dialog box, check **Analysis ToolPak** in the **Add-Ins Available** list and click **OK**.

If Analysis ToolPak does not appear in the list, rerun the Microsoft Office setup program to install this component if you use Microsoft Excel with Microsoft Windows.

APPENDIX

B

Review of Arithmetic and Algebra

The authors understand and realize that wide differences exist in the mathematical background of readers of this book. Some of you might have taken various courses in algebra, calculus, and matrix algebra, while others might not have taken any mathematics courses in a long period of time. Because the emphasis in this book is on statistical concepts and the interpretation of spreadsheet results, no prerequisite beyond elementary algebra is needed. To assess your arithmetic and algebraic skills, answer the following questions and then read the review that follows.

Assessment Quiz

Part 1

Fill in the correct answer.

1. $\dfrac{\frac{1}{2}}{\frac{}{3}} =$

2. $(0.4)^2 =$

3. $1 + \dfrac{2}{3} =$

4. $\left(\dfrac{1}{3}\right)^{(4)} =$

5. $\frac{1}{5} =$ (in decimals)

6. $1 - (-0.3) =$

7. $4 \times 0.2 \times (-8) =$

8. $\left(\frac{1}{4} \times \frac{2}{3}\right) =$

9. $\left(\frac{1}{100}\right) + \left(\frac{1}{200}\right) =$

10. $\sqrt{16} =$

Part 2

Select the correct answer.

1. If $a = bc$, then $c =$
   a. ab
   b. b/a
   c. a/b
   d. none of the above
2. If $x + y = z$, then $y =$
   a. z/x
   b. $z + x$
   c. $z - x$
   d. none of the above
3. $(x^3)(x^2) =$
   a. x^5
   b. x^6
   c. x^1
   d. none of the above
4. $x^0 =$
   a. x
   b. 1
   c. 0
   d. none of the above

5. $x(y - z) =$
   a. $xy - xz$
   b. $xy - z$
   c. $(y - z)/x$
   d. none of the above

6. $(x + y)/z =$
   a. $(x/z) + y$
   b. $(x/z) + (y/z)$
   c. $x + (y/z)$
   d. none of the above

7. $x/(y + z) =$
   a. $(x/y) + (1/z)$
   b. $(x/y) + (x/z)$
   c. $(y + z)/x$
   d. none of the above

8. If $x = 10$, $y = 5$, $z = 2$, and $w = 20$, then $(xy - z^2)/w =$
   a. 5
   b. 2.3
   c. 46
   d. none of the above

9. $(8x^4)/(4x^2) =$
   a. $2x^2$
   b. 2
   c. $2x$
   d. none of the above

10. $\sqrt{\dfrac{X}{Y}} =$
   a. $\sqrt{Y}/\sqrt{X}$
   b. $\sqrt{1}/\sqrt{XY}$
   c. $\sqrt{X}/\sqrt{Y}$
   d. none of the above

The answers to both parts of the quiz appear at the end of this appendix.

Symbols

Each of the four basic arithmetic operations—addition, subtraction, multiplication, and division—is indicated by a symbol.

$+$ add

$\times$ or $\cdot$ multiply

$-$ subtract

$\div$ or / divide

In addition to these operations, the following symbols are used to indicate equality or inequality:

$=$ equals

$\neq$ not equal

$\cong$ approximately equal to

$>$ greater than

$<$ less than

$\geq$ greater than or equal to

$\leq$ less than or equal to

Addition

Addition refers to the summation or accumulation of a set of numbers. In adding numbers, the two basic laws are the commutative law and the associative law.

The **commutative law** of addition states that the order in which numbers are added is irrelevant. This can be seen in the following two examples:

$1 + 2 = 3 \qquad 2 + 1 = 3$

$x + y = z \qquad y + x = z$

In each example, the number that was listed first and the number that was listed second did not matter.

The **associative law** of addition states that in adding several numbers, any subgrouping of the numbers can be added first, last, or in the middle. You can see this in the following examples:

$2 + 3 + 6 + 7 + 4 + 1 = 23$

$(5) + (6 + 7) + 4 + 1 = 23$

$5 + 13 + 5 = 23$

$5 + 6 + 7 + 4 + 1 = 23$

In each of these examples, the order in which the numbers have been added has no effect on the results.

Subtraction

The process of subtraction is the opposite or inverse of addition. The operation of subtracting 1 from 2 (that is, $2 - 1$) means that one unit is to be taken away from two units, leaving a remainder of one unit. In contrast to addition, the commutative and associative laws do not hold for subtraction. Therefore, as indicated in the following examples,

$8 - 4 = 4$	but	$4 - 8 = -4$
$3 - 6 = -3$	but	$6 - 3 = 3$
$8 - 3 - 2 = 3$	but	$3 - 2 - 8 = -7$
$9 - 4 - 2 = 3$	but	$2 - 4 - 9 = -11$

When subtracting negative numbers, remember that the same result occurs when subtracting a negative number as when adding a positive number. Thus,

$4 - (-3) = +7$	$4 + 3 = 7$
$8 - (-10) = +18$	$8 + 10 = 18$

Multiplication

The operation of multiplication is a shortcut method of addition when the same number is to be added several times. For example, if 7 is added three times ($7 + 7 + 7$), you could multiply 7 times 3 to get the product of 21.

In multiplication as in addition, the commutative laws and associative laws are in operation so that:

$$a \times b = b \times a$$

$$4 \times 5 = 5 \times 4 = 20$$

$$(2 \times 5) \times 6 = 10 \times 6 = 60$$

A third law of multiplication, the **distributive law**, applies to the multiplication of one number by the sum of several numbers. Here,

$$a(b + c) = ab + ac$$

$$2(3 + 4) = 2(7) = 2(3) + 2(4) = 14$$

The resulting product is the same regardless of whether b and c are summed and multiplied by a, or a is multiplied by b and by c and the two products are added together.

You also need to remember that when multiplying negative numbers, a negative number multiplied by a negative number equals a positive number. Thus,

$$(-a) \times (-b) = ab$$

$$(-5) \times (-4) = +20$$

Division

Just as subtraction is the opposite of addition, division is the opposite or inverse of multiplication. Division can be viewed as a shortcut to subtraction. When you divide 20 by 4, you are actually determining the number of times that 4 can be subtracted from 20. In general, however, the number of times one number can be divided by another may not be an exact integer value because there could be a remainder. For example, if you divide 21 by 4, the answer is 5 with a remainder of 1, or $5\frac{1}{4}$.

As in the case of subtraction, neither the commutative nor associative law of addition and multiplication holds for division.

$$a \div b \neq b \div a$$

$$9 \div 3 \neq 3 \div 9$$

$$6 \div (3 \div 2) = 4$$

$$(6 \div 3) \div 2 = 1$$

The distributive law holds only when the numbers to be added are contained in the numerator, not the denominator. Thus,

$$\frac{a+b}{c} = \frac{a}{c} + \frac{b}{c} \quad \text{but} \quad \frac{a}{b+c} \neq \frac{a}{b} + \frac{a}{c}$$

For example,

$$\frac{1}{2+3} = \frac{1}{5} \quad \text{but} \quad \frac{1}{2+3} \neq \frac{1}{2} + \frac{1}{3}$$

The last important property of division states that if the numerator and the denominator are multiplied or divided by the same number, the resulting quotient is not affected. Therefore,

$$\frac{80}{40} = 2$$

then

$$\frac{5(80)}{5(40)} = \frac{400}{200} = 2$$

and

$$\frac{80 \div 5}{40 \div 5} = \frac{16}{8} = 2$$

Fractions

A fraction is a number that consists of a combination of whole numbers and/or parts of whole numbers. For instance, the fraction 1/3 consists of only one portion of a number, while the fraction 7/6 consists of the whole number 1 plus the fraction 1/6. Each of the operations of addition, subtraction, multiplication, and division can be used with fractions. When adding and subtracting fractions, you must find the lowest common denominator for each fraction prior to adding or subtracting them. Thus, in adding 1/3 + 1/5, the lowest common denominator is 15, so

$$\frac{5}{15} + \frac{3}{15} = \frac{8}{15}$$

In subtracting 1/4 – 1/6, the same principles applies, so that the lowest common denominator is 12, producing a result of

$$\frac{3}{12} - \frac{2}{12} = \frac{1}{12}$$

Multiplying and dividing fractions does not have the lowest common denominator requirement associated with adding and subtracting fractions. Thus, if a/b is multiplied by c/d, the result is ac/bd.

The resulting numerator, *ac*, is the product of the numerators *a* and *c*, while the denominator, *bd*, is the product of the two denominators *b* and *d*. The resulting fraction can sometimes be reduced to a lower term by dividing the numerator and denominator by a common factor. For example, taking

$$\frac{2}{3} \times \frac{6}{7} = \frac{12}{21}$$

and dividing the numerator and denominator by 3 produces the result 4/7.

Division of fractions can be thought of as the inverse of multiplication, so the divisor can be inverted and multiplied by the original fraction. Thus,

$$\frac{9}{5} \div \frac{1}{4} = \frac{9}{5} \times \frac{4}{1} = \frac{36}{5}$$

The division of a fraction can also be thought of as a way of converting the fraction to a decimal number. For example, the fraction 2/5 can be converted to a decimal number by dividing its numerator, 2, by its denominator, 5, to produce the decimal number 0.40.

Exponents and Square Roots

Exponentiation (raising a number to a power) provides a shortcut in writing numerous multiplications. For example, $2 \times 2 \times 2 \times 2 \times 2$ can be written as $2^5 = 32$. The 5 represents the exponent (or power) of the number 2, telling you that 2 is to be multiplied by itself five times.

Several rules can be used for multiplying or dividing numbers that contain exponents.

Rule 1 $x^a \cdot x^b = x^{(a+b)}$

If two numbers involving a power of the same number are multiplied, the product is the same number raised to the sum of the powers.

$$4^2 \cdot 4^3 = (4 \cdot 4)(4 \cdot 4 \cdot 4) = 4^5$$

Rule 2 $(x^a)^b = x^{ab}$

If you take the power of a number that is already taken to a power, the result is a number that is raised to the product of the two powers. For example,

$$(4^2)^3 = (4^2)(4^2)(4^2) = 4^6$$

Rule 3 $\frac{x^a}{x^b} = x^{(a-b)}$

If a number raised to a power is divided by the same number raised to a power, the quotient is the number raised to the difference of the powers. Thus,

$$\frac{3^5}{3^3} = \frac{3 \cdot 3 \cdot 3 \cdot 3 \cdot 3}{3 \cdot 3 \cdot 3} = 3^2$$

If the denominator has a higher power than the numerator, the resulting quotient is a negative power. Thus,

$$\frac{3^3}{3^5} = \frac{3 \cdot 3 \cdot 3}{3 \cdot 3 \cdot 3 \cdot 3 \cdot 3} = \frac{1}{3^2} = 3^{-2} = \frac{1}{9}$$

If the difference between the powers of the numerator and denominator is 1, the result is the number itself. In other words, $x^1 = x$. For example,

$$\frac{3^3}{3^2} = \frac{3 \cdot 3 \cdot 3}{3 \cdot 3} = 3^1 = 3$$

If, however, no difference exists in the power of the numbers in the numerator and denominator, the result is 1. Thus,

$$\frac{x^a}{x^a} = x^{a-a} = x^0 = 1$$

Therefore, any number raised to the zero power equals 1. For example,

$$\frac{3^3}{3^3} = \frac{3 \cdot 3 \cdot 3}{3 \cdot 3 \cdot 3} = 3^0 = 1$$

The square root represented by the symbol $\sqrt{\ }$ is a special power of a number, the ½ power. It indicates the value that when multiplied by itself, will produce the original number.

Equations

In statistics, many formulas are expressed as equations where one unknown value is a function of another value. Thus, it is important that you know how to manipulate equations into various forms. The rules of addition, subtraction, multiplication, and division can be used to work with equations. For example, the equation

$$x - 2 = 5$$

can be solved for x by adding 2 to each side of the equation. This results in

$x - 2 + 2 = 5 + 2$. Therefore, $x = 7$.

If

$$x + y = z$$

you could solve for x by subtracting y from both sides of the equation so that

$x + y - y = z - y$. Therefore, $x = z - y$.

If the product of two variables is equal to a third variable, such as

$$x\,y = z,$$

you can solve for x by dividing both sides of the equation by y. Thus,

$$\frac{xy}{y} = \frac{z}{y}$$

$$x = \frac{z}{y}$$

Conversely, if

$$\frac{x}{y} = z$$

you can solve for x by multiplying both sides of the equation by y.

$$\frac{xy}{y} = zy$$

$$x = zy$$

To summarize, the various operations of addition, subtraction, multiplication, and division can be applied to equations as long as the same operation is performed on each side of the equation, thereby maintaining the equality.

Answers to Quiz

Part 1

1. $3/2$
2. 0.16
3. $5/3$
4. $1/81$
5. 0.20
6. 1.30
7. –6.4
8. $+1/6$
9. $3/200$
10. 4

Part 2

1. c
2. c
3. a
4. b
5. a
6. b
7. d
8. b
9. a
10. c

Statistical Tables

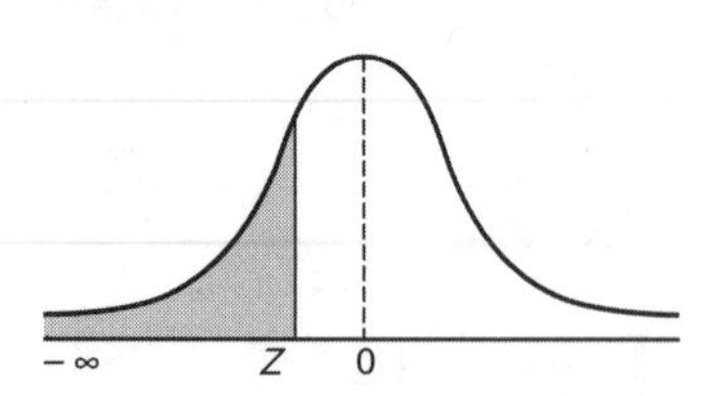

TABLE C.1

The Cumulative Standardized Normal Distribution

Entry represents area under the cumulative standardized normal distribution from −∞ to Z.

Z	0.00	0.01	0.02	0.03	0.04	0.05	0.06	0.07	0.08	0.09
−3.9	0.00005	0.00005	0.00004	0.00004	0.00004	0.00004	0.00004	0.00004	0.00003	0.00003
−3.8	0.00007	0.00007	0.00007	0.00006	0.00006	0.00006	0.00006	0.00005	0.00005	0.00005
−3.7	0.00011	0.00010	0.00010	0.00010	0.00009	0.00009	0.00008	0.00008	0.00008	0.00008
−3.6	0.00016	0.00015	0.00015	0.00014	0.00014	0.00013	0.00013	0.00012	0.00012	0.00011
−3.5	0.00023	0.00022	0.00022	0.00021	0.00020	0.00019	0.00019	0.00018	0.00017	0.00017
−3.4	0.00034	0.00032	0.00031	0.00030	0.00029	0.00028	0.00027	0.00026	0.00025	0.00024
−3.3	0.00048	0.00047	0.00045	0.00043	0.00042	0.00040	0.00039	0.00038	0.00036	0.00035
−3.2	0.00069	0.00066	0.00064	0.00062	0.00060	0.00058	0.00056	0.00054	0.00052	0.00050
−3.1	0.00097	0.00094	0.00090	0.00087	0.00084	0.00082	0.00079	0.00076	0.00074	0.00071
−3.0	0.00135	0.00131	0.00126	0.00122	0.00118	0.00114	0.00111	0.00107	0.00103	0.00100
−2.9	0.0019	0.0018	0.0018	0.0017	0.0016	0.0016	0.0015	0.0015	0.0014	0.0014
−2.8	0.0026	0.0025	0.0024	0.0023	0.0023	0.0022	0.0021	0.0021	0.0020	0.0019
−2.7	0.0035	0.0034	0.0033	0.0032	0.0031	0.0030	0.0029	0.0028	0.0027	0.0026
−2.6	0.0047	0.0045	0.0044	0.0043	0.0041	0.0040	0.0039	0.0038	0.0037	0.0036
−2.5	0.0062	0.0060	0.0059	0.0057	0.0055	0.0054	0.0052	0.0051	0.0049	0.0048
−2.4	0.0082	0.0080	0.0078	0.0075	0.0073	0.0071	0.0069	0.0068	0.0066	0.0064
−2.3	0.0107	0.0104	0.0102	0.0099	0.0096	0.0094	0.0091	0.0089	0.0087	0.0084
−2.2	0.0139	0.0136	0.0132	0.0129	0.0125	0.0122	0.0119	0.0116	0.0113	0.0110
−2.1	0.0179	0.0174	0.0170	0.0166	0.0162	0.0158	0.0154	0.0150	0.0146	0.0143
−2.0	0.0228	0.0222	0.0217	0.0212	0.0207	0.0202	0.0197	0.0192	0.0188	0.0183

Z	0.00	0.01	0.02	0.03	0.04	0.05	0.06	0.07	0.08	0.09
−1.9	0.0287	0.0281	0.0274	0.0268	0.0262	0.0256	0.0250	0.0244	0.0239	0.0233
−1.8	0.0359	0.0351	0.0344	0.0336	0.0329	0.0322	0.0314	0.0307	0.0301	0.0294
−1.7	0.0446	0.0436	0.0427	0.0418	0.0409	0.0401	0.0392	0.0384	0.0375	0.0367
−1.6	0.0548	0.0537	0.0526	0.0516	0.0505	0.0495	0.0485	0.0475	0.0465	0.0455
−1.5	0.0668	0.0655	0.0643	0.0630	0.0618	0.0606	0.0594	0.0582	0.0571	0.0559
−1.4	0.0808	0.0793	0.0778	0.0764	0.0749	0.0735	0.0721	0.0708	0.0694	0.0681
−1.3	0.0968	0.0951	0.0934	0.0918	0.0901	0.0885	0.0869	0.0853	0.0838	0.0823
−1.2	0.1151	0.1131	0.1112	0.1093	0.1075	0.1056	0.1038	0.1020	0.1003	0.0985
−1.1	0.1357	0.1335	0.1314	0.1292	0.1271	0.1251	0.1230	0.1210	0.1190	0.1170
−1.0	0.1587	0.1562	0.1539	0.1515	0.1492	0.1469	0.1446	0.1423	0.1401	0.1379
−0.9	0.1841	0.1814	0.1788	0.1762	0.1736	0.1711	0.1685	0.1660	0.1635	0.1611
−0.8	0.2119	0.2090	0.2061	0.2033	0.2005	0.1977	0.1949	0.1922	0.1894	0.1867
−0.7	0.2420	0.2388	0.2358	0.2327	0.2296	0.2266	0.2236	0.2206	0.2177	0.2148
−0.6	0.2743	0.2709	0.2676	0.2643	0.2611	0.2578	0.2546	0.2514	0.2482	0.2451
−0.5	0.3085	0.3050	0.3015	0.2981	0.2946	0.2912	0.2877	0.2843	0.2810	0.2776
−0.4	0.3446	0.3409	0.3372	0.3336	0.3300	0.3264	0.3228	0.3192	0.3156	0.3121
−0.3	0.3821	0.3783	0.3745	0.3707	0.3669	0.3632	0.3594	0.3557	0.3520	0.3483
−0.2	0.4207	0.4168	0.4129	0.4090	0.4052	0.4013	0.3974	0.3936	0.3897	0.3859
−0.1	0.4602	0.4562	0.4522	0.4483	0.4443	0.4404	0.4364	0.4325	0.4286	0.4247
−0.0	0.5000	0.4960	0.4920	0.4880	0.4840	0.4801	0.4761	0.4721	0.4681	0.4641

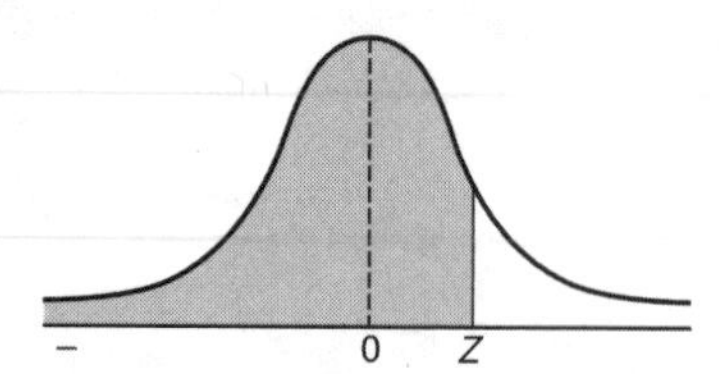

TABLE C.1 Continued

Entry represents area under the standardized normal distribution from $-\infty$ to Z.

Z	0.00	0.01	0.02	0.03	0.04	0.05	0.06	0.07	0.08	0.09
0.0	0.5000	0.5040	0.5080	0.5120	0.5160	0.5199	0.5239	0.5279	0.5319	0.5359
0.1	0.5398	0.5438	0.5478	0.5517	0.5557	0.5596	0.5636	0.5675	0.5714	0.5753
0.2	0.5793	0.5832	0.5871	0.5910	0.5948	0.5987	0.6026	0.6064	0.6103	0.6141
0.3	0.6179	0.6217	0.6255	0.6293	0.6331	0.6368	0.6406	0.6443	0.6480	0.6517
0.4	0.6554	0.6591	0.6628	0.6664	0.6700	0.6736	0.6772	0.6808	0.6844	0.6879
0.5	0.6915	0.6950	0.6985	0.7019	0.7054	0.7088	0.7123	0.7157	0.7190	0.7224
0.6	0.7257	0.7291	0.7324	0.7357	0.7389	0.7422	0.7454	0.7486	0.7518	0.7549
0.7	0.7580	0.7612	0.7642	0.7673	0.7704	0.7734	0.7764	0.7794	0.7823	0.7852
0.8	0.7881	0.7910	0.7939	0.7967	0.7995	0.8023	0.8051	0.8078	0.8106	0.8133
0.9	0.8159	0.8186	0.8212	0.8238	0.8264	0.8289	0.8315	0.8340	0.8365	0.8389
1.0	0.8413	0.8438	0.8461	0.8485	0.8508	0.8531	0.8554	0.8577	0.8599	0.8621
1.1	0.8643	0.8665	0.8686	0.8708	0.8729	0.8749	0.8770	0.8790	0.8810	0.8830
1.2	0.8849	0.8869	0.8888	0.8907	0.8925	0.8944	0.8962	0.8980	0.8997	0.9015
1.3	0.9032	0.9049	0.9066	0.9082	0.9099	0.9115	0.9131	0.9147	0.9162	0.9177
1.4	0.9192	0.9207	0.9222	0.9236	0.9251	0.9265	0.9279	0.9292	0.9306	0.9319
1.5	0.9332	0.9345	0.9357	0.9370	0.9382	0.9394	0.9406	0.9418	0.9429	0.9441
1.6	0.9452	0.9463	0.9474	0.9484	0.9495	0.9505	0.9515	0.9525	0.9535	0.9545
1.7	0.9554	0.9564	0.9573	0.9582	0.9591	0.9599	0.9608	0.9616	0.9625	0.9633
1.8	0.9641	0.9649	0.9656	0.9664	0.9671	0.9678	0.9686	0.9693	0.9699	0.9706
1.9	0.9713	0.9719	0.9726	0.9732	0.9738	0.9744	0.9750	0.9756	0.9761	0.9767
2.0	0.9772	0.9778	0.9783	0.9788	0.9793	0.9798	0.9803	0.9808	0.9812	0.9817
2.1	0.9821	0.9826	0.9830	0.9834	0.9838	0.9842	0.9846	0.9850	0.9854	0.9857
2.2	0.9861	0.9864	0.9868	0.9871	0.9875	0.9878	0.9881	0.9884	0.9887	0.9890

Z	0.00	0.01	0.02	0.03	0.04	0.05	0.06	0.07	0.08	0.09
2.3	0.9893	0.9896	0.9898	0.9901	0.9904	0.9906	0.9909	0.9911	0.9913	0.9916
2.4	0.9918	0.9920	0.9922	0.9925	0.9927	0.9929	0.9931	0.9932	0.9934	0.9936
2.5	0.9938	0.9940	0.9941	0.9943	0.9945	0.9946	0.9948	0.9949	0.9951	0.9952
2.6	0.9953	0.9955	0.9956	0.9957	0.9959	0.9960	0.9961	0.9962	0.9963	0.9964
2.7	0.9965	0.9966	0.9967	0.9968	0.9969	0.9970	0.9971	0.9972	0.9973	0.9974
2.8	0.9974	0.9975	0.9976	0.9977	0.9977	0.9978	0.9979	0.9979	0.9980	0.9981
2.9	0.9981	0.9982	0.9982	0.9983	0.9984	0.9984	0.9985	0.9985	0.9986	0.9986
3.0	0.99865	0.99869	0.99874	0.99878	0.99882	0.99886	0.99889	0.99893	0.99897	0.99900
3.1	0.99903	0.99906	0.99910	0.99913	0.99916	0.99918	0.99921	0.99924	0.99926	0.99929
3.2	0.99931	0.99934	0.99936	0.99938	0.99940	0.99942	0.99944	0.99946	0.99948	0.99950
3.3	0.99952	0.99953	0.99955	0.99957	0.99958	0.99960	0.99961	0.99962	0.99964	0.99965
3.4	0.99966	0.99968	0.99969	0.99970	0.99971	0.99972	0.99973	0.99974	0.99975	0.99976
3.5	0.99977	0.99978	0.99978	0.99979	0.99980	0.99981	0.99981	0.99982	0.99983	0.99983
3.6	0.99984	0.99985	0.99985	0.99986	0.99986	0.99987	0.99987	0.99988	0.99988	0.99989
3.7	0.99989	0.99990	0.99990	0.99990	0.99991	0.99991	0.99992	0.99992	0.99992	0.99992
3.8	0.99993	0.99993	0.99993	0.99994	0.99994	0.99994	0.99994	0.99995	0.99995	0.99995
3.9	0.99995	0.99995	0.99996	0.99996	0.99996	0.99996	0.99996	0.99996	0.99997	0.99997
4.0	0.99996832									
4.5	0.99999660									
5.0	0.99999971									
5.5	0.99999998									
6.0	0.99999999									

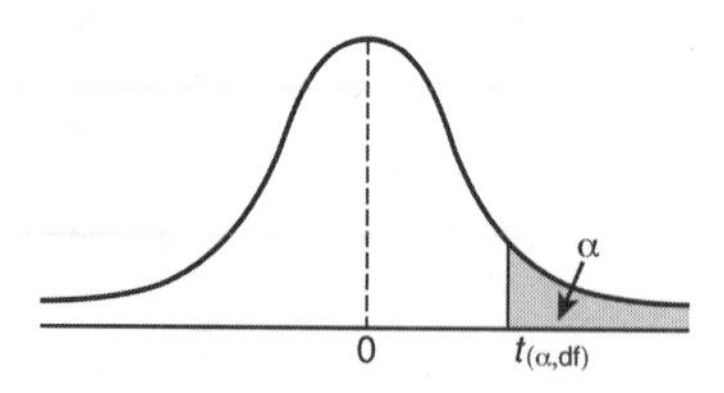

TABLE C.2

Critical Values of t

	Upper-Tail Areas					
Degrees of Freedom	**0.25**	**0.10**	**0.05**	**0.025**	**0.01**	**0.005**
1	1.0000	3.0777	6.3138	12.7062	31.8207	63.6574
2	0.8165	1.8856	2.9200	4.3027	6.9646	9.9248
3	0.7649	1.6377	2.3534	3.1824	4.5407	5.8409
4	0.7407	1.5332	2.1318	2.7764	3.7469	4.6041
5	0.7267	1.4759	2.0150	2.5706	3.3649	4.0322
6	0.7176	1.4398	1.9432	2.4469	3.1427	3.7074
7	0.7111	1.4149	1.8946	2.3646	2.9980	3.4995
8	0.7064	1.3968	1.8595	2.3060	2.8965	3.3554
9	0.7027	1.3830	1.8331	2.2622	2.8214	3.2498
10	0.6998	1.3722	1.8125	2.2281	2.7638	3.1693
11	0.6974	1.3634	1.7959	2.2010	2.7181	3.1058
12	0.6955	1.3562	1.7823	2.1788	2.6810	3.0545
13	0.6938	1.3502	1.7709	2.1604	2.6503	3.0123
14	0.6924	1.3450	1.7613	2.1448	2.6245	2.9768
15	0.6912	1.3406	1.7531	2.1315	2.6025	2.9467
16	0.6901	1.3368	1.7459	2.1199	2.5835	2.9208
17	0.6892	1.3334	1.7396	2.1098	2.5669	2.8982
18	0.6884	1.3304	1.7341	2.1009	2.5524	2.8784
19	0.6876	1.3277	1.7291	2.0930	2.5395	2.8609
20	0.6870	1.3253	1.7247	2.0860	2.5280	2.8453
21	0.6864	1.3232	1.7207	2.0796	2.5177	2.8314
22	0.6858	1.3212	1.7171	2.0739	2.5083	2.8188
23	0.6853	1.3195	1.7139	2.0687	2.4999	2.8073
24	0.6848	1.3178	1.7109	2.0639	2.4922	2.7969
25	0.6844	1.3163	1.7081	2.0595	2.4851	2.7874
26	0.6840	1.3150	1.7056	2.0555	2.4786	2.7787

Degrees of Freedom	Upper-Tail Areas					
	0.25	0.10	0.05	0.025	0.01	0.005
27	0.6837	1.3137	1.7033	2.0518	2.4727	2.7707
28	0.6834	1.3125	1.7011	2.0484	2.4671	2.7633
29	0.6830	1.3114	1.6991	2.0452	2.4620	2.7564
30	0.6828	1.3104	1.6973	2.0423	2.4573	2.7500
31	0.6825	1.3095	1.6955	2.0395	2.4528	2.7440
32	0.6822	1.3086	1.6939	2.0369	2.4487	2.7385
33	0.6820	1.3077	1.6924	2.0345	2.4448	2.7333
34	0.6818	1.3070	1.6909	2.0322	2.4411	2.7284
35	0.6816	1.3062	1.6896	2.0301	2.4377	2.7238
36	0.6814	1.3055	1.6883	2.0281	2.4345	2.7195
37	0.6812	1.3049	1.6871	2.0262	2.4314	2.7154
38	0.6810	1.3042	1.6860	2.0244	2.4286	2.7116
39	0.6808	1.3036	1.6849	2.0227	2.4258	2.7079
40	0.6807	1.3031	1.6839	2.0211	2.4233	2.7045
41	0.6805	1.3025	1.6829	2.0195	2.4208	2.7012
42	0.6804	1.3020	1.6820	2.0181	2.4185	2.6981
43	0.6802	1.3016	1.6811	2.0167	2.4163	2.6951
44	0.6801	1.3011	1.6802	2.0154	2.4141	2.6923
45	0.6800	1.3006	1.6794	2.0141	2.4121	2.6896
46	0.6799	1.3022	1.6787	2.0129	2.4102	2.6870
47	0.6797	1.2998	1.6779	2.0117	2.4083	2.6846
48	0.6796	1.2994	1.6772	2.0106	2.4066	2.6822
49	0.6795	1.2991	1.6766	2.0096	2.4049	2.6800
50	0.6794	1.2987	1.6759	2.0086	2.4033	2.6778
51	0.6793	1.2984	1.6753	2.0076	2.4017	2.6757

(continues)

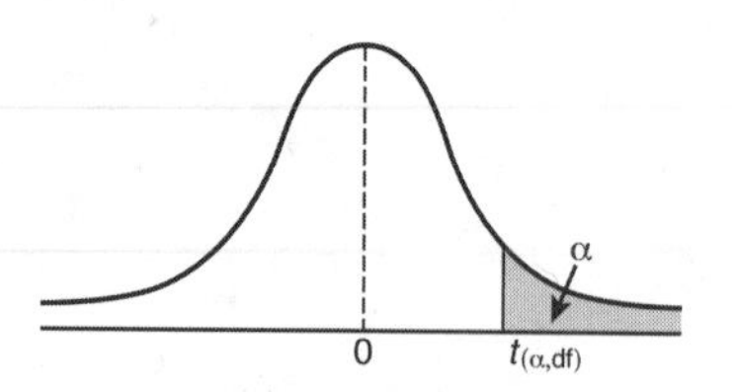

TABLE C.2 Continued

	Upper-Tail Areas					
Degrees of Freedom	**0.25**	**0.10**	**0.05**	**0.025**	**0.01**	**0.005**
52	0.6792	1.2980	1.6747	2.0066	2.4002	2.6737
53	0.6791	1.2977	1.6741	2.0057	2.3988	2.6718
54	0.6791	1.2974	1.6736	2.0049	2.3974	2.6700
55	0.6790	1.2971	1.6730	2.0040	2.3961	2.6682
56	0.6789	1.2969	1.6725	2.0032	2.3948	2.6665
57	0.6788	1.2966	1.6720	2.0025	2.3936	2.6649
58	0.6787	1.2963	1.6716	2.0017	2.3924	2.6633
59	0.6787	1.2961	1.6711	2.0010	2.3912	2.6618
60	0.6786	1.2958	1.6706	2.0003	2.3901	2.6603
61	0.6785	1.2956	1.6702	1.9996	2.3890	2.6589
62	0.6785	1.2954	1.6698	1.9990	2.3880	2.6575
63	0.6784	1.2951	1.6694	1.9983	2.3870	2.6561
64	0.6783	1.2949	1.6690	1.9977	2.3860	2.6549
65	0.6783	1.2947	1.6686	1.9971	2.3851	2.6536
66	0.6782	1.2945	1.6683	1.9966	2.3842	2.6524
67	0.6782	1.2943	1.6679	1.9960	2.3833	2.6512
68	0.6781	1.2941	1.6676	1.9955	2.3824	2.6501
69	0.6781	1.2939	1.6672	1.9949	2.3816	2.6490
70	0.6780	1.2938	1.6669	1.9944	2.3808	2.6479
71	0.6780	1.2936	1.6666	1.9939	2.3800	2.6469
72	0.6779	1.2934	1.6663	1.9935	2.3793	2.6459
73	0.6779	1.2933	1.6660	1.9930	2.3785	2.6449
74	0.6778	1.2931	1.6657	1.9925	2.3778	2.6439
75	0.6778	1.2929	1.6654	1.9921	2.3771	2.6430
76	0.6777	1.2928	1.6652	1.9917	2.3764	2.6421
77	0.6777	1.2926	1.6649	1.9913	2.3758	2.6412

	Upper-Tail Areas					
Degrees of Freedom	**0.25**	**0.10**	**0.05**	**0.025**	**0.01**	**0.005**
78	0.6776	1.2925	1.6646	1.9908	2.3751	2.6403
79	0.6776	1.2924	1.6644	1.9905	2.3745	2.6395
80	0.6776	1.2922	1.6641	1.9901	2.3739	2.6387
81	0.6775	1.2921	1.6639	1.9897	2.3733	2.6379
82	0.6775	1.2920	1.6636	1.9893	2.3727	2.6371
83	0.6775	1.2918	1.6634	1.9890	2.3721	2.6364
84	0.6774	1.2917	1.6632	1.9886	2.3716	2.6356
85	0.6774	1.2916	1.6630	1.9883	2.3710	2.6349
86	0.6774	1.2915	1.6628	1.9879	2.3705	2.6342
87	0.6773	1.2914	1.6626	1.9876	2.3700	2.6335
88	0.6773	1.2912	1.6624	1.9873	2.3695	2.6329
89	0.6773	1.2911	1.6622	1.9870	2.3690	2.6322
90	0.6772	1.2910	1.6620	1.9867	2.3685	2.6316
91	0.6772	1.2909	1.6618	1.9864	2.3680	2.6309
92	0.6772	1.2908	1.6616	1.9861	2.3676	2.6303
93	0.6771	1.2907	1.6614	1.9858	2.3671	2.6297
94	0.6771	1.2906	1.6612	1.9855	2.3667	2.6291
95	0.6771	1.2905	1.6611	1.9853	2.3662	2.6286
96	0.6771	1.2904	1.6609	1.9850	2.3658	2.6280
97	0.6770	1.2903	1.6607	1.9847	2.3654	2.6275
98	0.6770	1.2902	1.6606	1.9845	2.3650	2.6269
99	0.6770	1.2902	1.6604	1.9842	2.3646	2.6264
100	0.6770	1.2901	1.6602	1.9840	2.3642	2.6259
110	0.6767	1.2893	1.6588	1.9818	2.3607	2.6213
120	0.6765	1.2886	1.6577	1.9799	2.3578	2.6174
∞	0.6745	1.2816	1.6449	1.9600	2.3263	2.5758

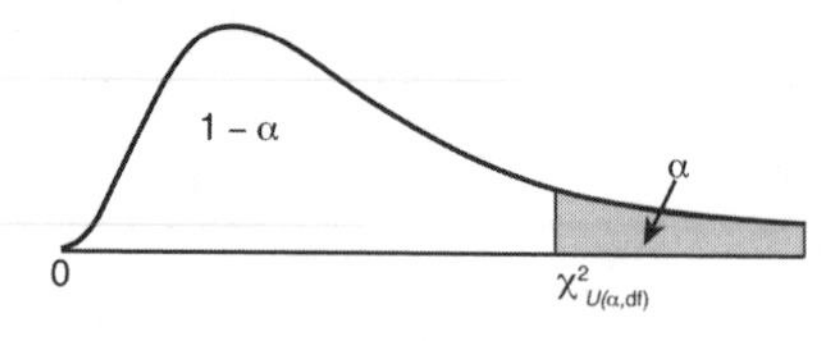

TABLE C.3

Critical Values of χ^2

For a particular number of degrees of freedom, entry represents the critical value of χ^2 corresponding to a specified upper-tail area (α).

	Upper Tail Areas (α)					
Degrees of Freedom	**0.995**	**0.99**	**0.975**	**0.95**	**0.90**	**0.75**
1			0.001	0.004	0.016	0.102
2	0.010	0.020	0.051	0.103	0.211	0.575
3	0.072	0.115	0.216	0.352	0.584	1.213
4	0.207	0.297	0.484	0.711	1.064	1.923
5	0.412	0.554	0.831	1.145	1.610	2.675
6	0.676	0.872	1.237	1.635	2.204	3.455
7	0.989	1.239	1.690	2.167	2.833	4.255
8	1.344	1.646	2.180	2.733	3.490	5.071
9	1.735	2.088	2.700	3.325	4.168	5.899
10	2.156	2.558	3.247	3.940	4.865	6.737
11	2.603	3.053	3.816	4.575	5.578	7.584
12	3.074	3.571	4.404	5.226	6.304	8.438
13	3.565	4.107	5.009	5.892	7.042	9.299
14	4.075	4.660	5.629	6.571	7.790	10.165
15	4.601	5.229	6.262	7.261	8.547	11.037
16	5.142	5.812	6.908	7.962	9.312	11.912
17	5.697	6.408	7.564	8.672	10.085	12.792
18	6.265	7.015	8.231	9.390	10.865	13.675
19	6.844	7.633	8.907	10.117	11.651	14.562
20	7.434	8.260	9.591	10.851	12.443	15.452
21	8.034	8.897	10.283	11.591	13.240	16.344
22	8.643	9.542	10.982	12.338	14.042	17.240
23	9.260	10.196	11.689	13.091	14.848	18.137
24	9.886	10.856	12.401	13.848	15.659	19.037
25	10.520	11.524	13.120	14.611	16.473	19.939
26	11.160	12.198	13.844	15.379	17.292	20.843
27	11.808	12.879	14.573	16.151	18.114	21.749
28	12.461	13.565	15.308	16.928	18.939	22.657
29	13.121	14.257	16.047	17.708	19.768	23.567
30	13.787	14.954	16.791	18.493	20.599	24.478

For larger values of degrees of freedom (df), the expression $Z = \sqrt{2\chi^2} - \sqrt{2(df)-1}$ may be used and the resulting upper-tail area can be obtained from the cumulative standardized normal distribution (Table C.1).

Upper Tail Areas (α)					
0.25	0.10	0.05	0.025	0.01	0.005
1.323	2.706	3.841	5.024	6.635	7.879
2.773	4.605	5.991	7.378	9.210	10.597
4.108	6.251	7.815	9.348	11.345	12.838
5.385	7.779	9.488	11.143	13.277	14.860
6.626	9.236	11.071	12.833	15.086	16.750
7.841	10.645	12.592	14.449	16.812	18.458
9.037	12.017	14.067	16.013	18.475	20.278
10.219	13.362	15.507	17.535	20.090	21.955
11.389	14.684	16.919	19.023	21.666	23.589
12.549	15.987	18.307	20.483	23.209	25.188
13.701	17.275	19.675	21.920	24.725	26.757
14.845	18.549	21.026	23.337	26.217	28.299
15.984	19.812	22.362	24.736	27.688	29.819
17.117	21.064	23.685	26.119	29.141	31.319
18.245	22.307	24.996	27.488	30.578	32.801
19.369	23.542	26.296	28.845	32.000	34.267
20.489	24.769	27.587	30.191	33.409	35.718
21.605	25.989	28.869	31.526	34.805	37.156
22.718	27.204	30.144	32.852	36.191	38.582
23.828	28.412	31.410	34.170	37.566	39.997
24.935	29.615	32.671	35.479	38.932	41.401
26.039	30.813	33.924	36.781	40.289	42.796
27.141	32.007	35.172	38.076	41.638	44.181
28.241	33.196	36.415	39.364	42.980	45.559
29.339	34.382	37.652	40.646	44.314	46.928
30.435	35.563	38.885	41.923	45.642	48.290
31.528	36.741	40.113	43.194	46.963	49.645
32.620	37.916	41.337	44.461	48.278	50.993
33.711	39.087	42.557	45.722	49.588	52.336
34.800	40.256	43.773	46.979	50.892	53.672

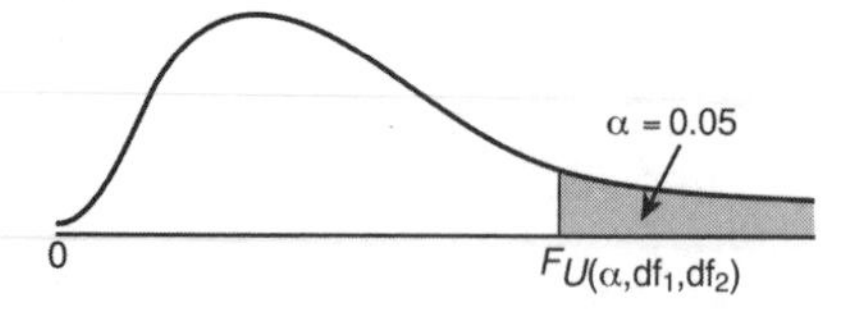

TABLE C.4

Critical Values of F

For a particular combination of numerator and denominator degrees of freedom, entry represents the critical values of *F* corresponding to a specified upper-tail area (α).

Denominator,	Numerator, df_1								
df_2	1	2	3	4	5	6	7	8	9
1	161.40	199.50	215.70	224.60	230.20	234.00	236.80	238.90	240.50
2	18.51	19.00	19.16	19.25	19.30	19.33	19.35	19.37	19.38
3	10.13	9.55	9.28	9.12	9.01	8.94	8.89	8.85	8.81
4	7.71	6.94	6.59	6.39	6.26	6.16	6.09	6.04	6.00
5	6.61	5.79	5.41	5.19	5.05	4.95	4.88	4.82	4.77
6	5.99	5.14	4.76	4.53	4.39	4.28	4.21	4.15	4.10
7	5.59	4.74	4.35	4.12	3.97	3.87	3.79	3.73	3.68
8	5.32	4.46	4.07	3.84	3.69	3.58	3.50	3.44	3.39
9	5.12	4.26	3.86	3.63	3.48	3.37	3.29	3.23	3.18
10	4.96	4.10	3.71	3.48	3.33	3.22	3.14	3.07	3.02
11	4.84	3.98	3.59	3.36	3.20	3.09	3.01	2.95	2.90
12	4.75	3.89	3.49	3.26	3.11	3.00	2.91	2.85	2.80
13	4.67	3.81	3.41	3.18	3.03	2.92	2.83	2.77	2.71
14	4.60	3.74	3.34	3.11	2.96	2.85	2.76	2.70	2.65
15	4.54	3.68	3.29	3.06	2.90	2.79	2.71	2.64	2.59
16	4.49	3.63	3.24	3.01	2.85	2.74	2.66	2.59	2.54
17	4.45	3.59	3.20	2.96	2.81	2.70	2.61	2.55	2.49
18	4.41	3.55	3.16	2.93	2.77	2.66	2.58	2.51	2.46
19	4.38	3.52	3.13	2.90	2.74	2.63	2.54	2.48	2.42

Numerator, df_1									
10	**12**	**15**	**20**	**24**	**30**	**40**	**60**	**120**	**∞**
241.90	243.90	245.90	248.00	249.10	250.10	251.10	252.20	253.30	254.30
19.40	19.41	19.43	19.45	19.45	19.46	19.47	19.48	19.49	19.50
8.79	8.74	8.70	8.66	8.64	8.62	8.59	8.57	8.55	8.53
5.96	5.91	5.86	5.80	5.77	5.75	5.72	5.69	5.66	5.63
4.74	4.68	4.62	4.56	4.53	4.50	4.46	4.43	4.40	4.36
4.06	4.00	3.94	3.87	3.84	3.81	3.77	3.74	3.70	3.67
3.64	3.57	3.51	3.44	3.41	3.38	3.34	3.30	3.27	3.23
3.35	3.28	3.22	3.15	3.12	3.08	3.04	3.01	2.97	2.93
3.14	3.07	3.01	2.94	2.90	2.86	2.83	2.79	2.75	2.71
2.98	2.91	2.85	2.77	2.74	2.70	2.66	2.62	2.58	2.54
2.85	2.79	2.72	2.65	2.61	2.57	2.53	2.49	2.45	2.40
2.75	2.69	2.62	2.54	2.51	2.47	2.43	2.38	2.34	2.30
2.67	2.60	2.53	2.46	2.42	2.38	2.34	2.30	2.25	2.21
2.60	2.53	2.46	2.39	2.35	2.31	2.27	2.22	2.18	2.13
2.54	2.48	2.40	2.33	2.29	2.25	2.20	2.16	2.11	2.07
2.49	2.42	2.35	2.28	2.24	2.19	2.15	2.11	2.06	2.01
2.45	2.38	2.31	2.23	2.19	2.15	2.10	2.06	2.01	1.96
2.41	2.34	2.27	2.19	2.15	2.11	2.06	2.02	1.97	1.92
2.38	2.31	2.23	2.16	2.11	2.07	2.03	1.98	1.93	1.88

(continues)

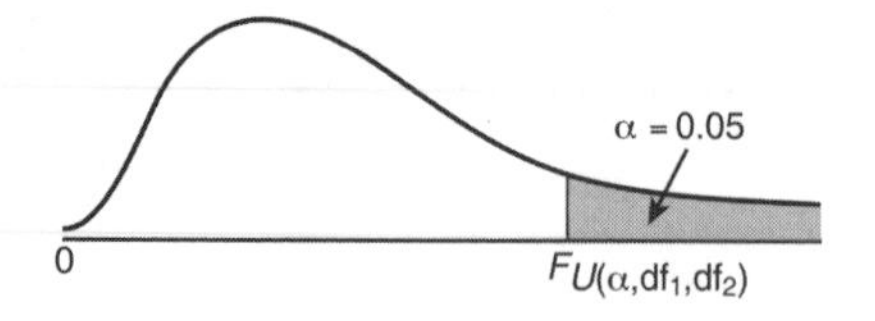

TABLE C.4 Continued

For a particular combination of numerator and denominator degrees of freedom, entry represents the critical values of F corresponding to a specified upper-tail area (α).

Denominator, df_2	Numerator, df_1								
	1	2	3	4	5	6	7	8	9
20	4.35	3.49	3.10	2.87	2.71	2.60	2.51	2.45	2.39
21	4.32	3.47	3.07	2.84	2.68	2.57	2.49	2.42	2.37
22	4.30	3.44	3.05	2.82	2.66	2.55	2.46	2.40	2.34
23	4.28	3.42	3.03	2.80	2.64	2.53	2.44	2.37	2.32
24	4.26	3.40	3.01	2.78	2.62	2.51	2.42	2.36	2.30
25	4.24	3.39	2.99	2.76	2.60	2.49	2.40	2.34	2.28
26	4.23	3.37	2.98	2.74	2.59	2.47	2.39	2.32	2.27
27	4.21	3.35	2.96	2.73	2.57	2.46	2.37	2.31	2.25
28	4.20	3.34	2.95	2.71	2.56	2.45	2.36	2.29	2.24
29	4.18	3.33	2.93	2.70	2.55	2.43	2.35	2.28	2.22
30	4.17	3.32	2.92	2.69	2.53	2.42	2.33	2.27	2.21
40	4.08	3.23	2.84	2.61	2.45	2.34	2.25	2.18	2.12
60	4.00	3.15	2.76	2.53	2.37	2.25	2.17	2.10	2.04
120	3.92	3.07	2.68	2.45	2.29	2.17	2.09	2.02	1.96
∞	3.84	3.00	2.60	2.37	2.21	2.10	2.01	1.94	1.88

Numerator, df_1									
10	12	15	20	24	30	40	60	120	∞
2.35	2.28	2.20	2.12	2.08	2.04	1.99	1.95	1.90	1.84
2.32	2.25	2.18	2.10	2.05	2.01	1.96	1.92	1.87	1.81
2.30	2.23	2.15	2.07	2.03	1.98	1.91	1.89	1.84	1.78
2.27	2.20	2.13	2.05	2.01	1.96	1.91	1.86	1.81	1.76
2.25	2.18	2.11	2.03	1.98	1.94	1.89	1.84	1.79	1.73
2.24	2.16	2.09	2.01	1.96	1.92	1.87	1.82	1.77	1.71
2.22	2.15	2.07	1.99	1.95	1.90	1.85	1.80	1.75	1.69
2.20	2.13	2.06	1.97	1.93	1.88	1.84	1.79	1.73	1.67
2.19	2.12	2.04	1.96	1.91	1.87	1.82	1.77	1.71	1.65
2.18	2.10	2.03	1.94	1.90	1.85	1.81	1.75	1.70	1.64
2.16	2.09	2.01	1.93	1.89	1.84	1.79	1.74	1.68	1.62
2.08	2.00	1.92	1.84	1.79	1.74	1.69	1.64	1.58	1.51
1.99	1.92	1.84	1.75	1.70	1.65	1.59	1.53	1.47	1.39
1.91	1.83	1.75	1.66	1.61	1.55	1.50	1.43	1.35	1.25
1.83	1.75	1.67	1.57	1.52	1.46	1.39	1.32	1.22	1.00

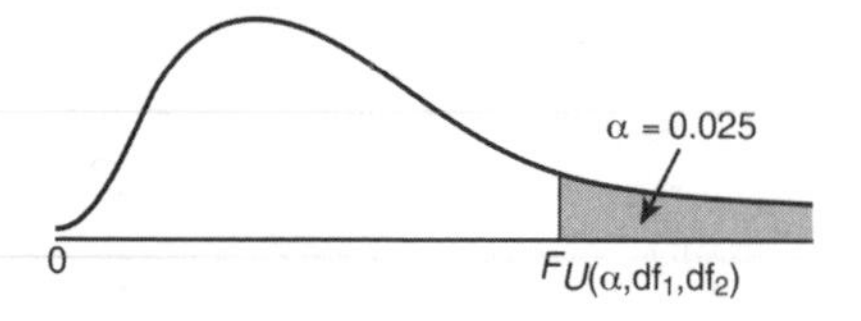

TABLE C.4 Continued

For a particular combination of numerator and denominator degrees of freedom, entry represents the critical values of F corresponding to a specified upper-tail area (α).

	Numerator, df_1								
Denominator, df_2	1	2	3	4	5	6	7	8	9
1	647.80	799.50	864.20	899.60	921.80	937.10	948.20	956.70	963.30
2	38.51	39.00	39.17	39.25	39.30	39.33	39.36	39.39	39.39
3	17.44	16.04	15.44	15.10	14.88	14.73	14.62	14.54	14.47
4	12.22	10.65	9.98	9.60	9.36	9.20	9.07	8.98	8.90
5	10.01	8.43	7.76	7.39	7.15	6.98	6.85	6.76	6.68
6	8.81	7.26	6.60	6.23	5.99	5.82	5.70	5.60	5.52
7	8.07	6.54	5.89	5.52	5.29	5.12	4.99	4.90	4.82
8	7.57	6.06	5.42	5.05	4.82	4.65	4.53	4.43	4.36
9	7.21	5.71	5.08	4.72	4.48	4.32	4.20	4.10	4.03
10	6.94	5.46	4.83	4.47	4.24	4.07	3.95	3.85	3.78
11	6.72	5.26	4.63	4.28	4.04	3.88	3.76	3.66	3.59
12	6.55	5.10	4.47	4.12	3.89	3.73	3.61	3.51	3.44
13	6.41	4.97	4.35	4.00	3.77	3.60	3.48	3.39	3.31
14	6.30	4.86	4.24	3.89	3.66	3.50	3.38	3.29	3.21
15	6.20	4.77	4.15	3.80	3.58	3.41	3.29	3.20	3.12
16	6.12	4.69	4.08	3.73	3.50	3.34	3.22	3.12	3.05
17	6.04	4.62	4.01	3.66	3.44	3.28	3.16	3.06	2.98
18	5.98	4.56	3.95	3.61	3.38	3.22	3.10	3.01	2.93
19	5.92	4.51	3.90	3.56	3.33	3.17	3.05	2.96	2.88
20	5.87	4.46	3.86	3.51	3.29	3.13	3.01	2.91	2.84
21	5.83	4.42	3.82	3.48	3.25	3.09	2.97	2.87	2.80

Numerator, df_1									
10	12	15	20	24	30	40	60	120	∞
968.60	976.70	984.90	993.10	997.20	1,001.00	1,006.00	1,010.00	1,014.00	1,018.00
39.40	39.41	39.43	39.45	39.46	39.46	39.47	39.48	39.49	39.50
14.42	14.34	14.25	14.17	14.12	14.08	14.04	13.99	13.95	13.90
8.84	8.75	8.66	8.56	8.51	8.46	8.41	8.36	8.31	8.26
6.62	6.52	6.43	6.33	6.28	6.23	6.18	6.12	6.07	6.02
5.46	5.37	5.27	5.17	5.12	5.07	5.01	4.96	4.90	4.85
4.76	4.67	4.57	4.47	4.42	4.36	4.31	4.25	4.20	4.14
4.30	4.20	4.10	4.00	3.95	3.89	3.84	3.78	3.73	3.67
3.96	3.87	3.77	3.67	3.61	3.56	3.51	3.45	3.39	3.33
3.72	3.62	3.52	3.42	3.37	3.31	3.26	3.20	3.14	3.08
3.53	3.43	3.33	3.23	3.17	3.12	3.06	3.00	2.94	2.88
3.37	3.28	3.18	3.07	3.02	2.96	2.91	2.85	2.79	2.72
3.25	3.15	3.05	2.95	2.89	2.84	2.78	2.72	2.66	2.60
3.15	3.05	2.95	2.84	2.79	2.73	2.67	2.61	2.55	2.49
3.06	2.96	2.86	2.76	2.70	2.64	2.59	2.52	2.46	2.40
2.99	2.89	2.79	2.68	2.63	2.57	2.51	2.45	2.38	2.32
2.92	2.82	2.72	2.62	2.56	2.50	2.44	2.38	2.32	2.25
2.87	2.77	2.67	2.56	2.50	2.44	2.38	2.32	2.26	2.19
2.82	2.72	2.62	2.51	2.45	2.39	2.33	2.27	2.20	2.13
2.77	2.68	2.57	2.46	2.41	2.35	2.29	2.22	2.16	2.09
2.73	2.64	2.53	2.42	2.37	2.31	2.25	2.18	2.11	2.04

(continues)

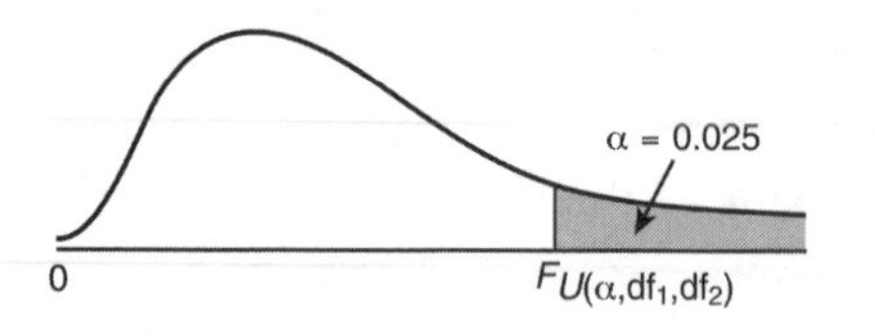

TABLE C.4 Continued

For a particular combination of numerator and denominator degrees of freedom, entry represents the critical values of *F* corresponding to a specified upper-tail area (α).

Denominator, df_2	Numerator, df_1 1	2	3	4	5	6	7	8	9
22	5.79	4.38	3.78	3.44	3.22	3.05	2.93	2.84	2.76
23	5.75	4.35	3.75	3.41	3.18	3.02	2.90	2.81	2.73
24	5.72	4.32	3.72	3.38	3.15	2.99	2.87	2.78	2.70
25	5.69	4.29	3.69	3.35	3.13	2.97	2.85	2.75	2.68
26	5.66	4.27	3.67	3.33	3.10	2.94	2.82	2.73	2.65
27	5.63	4.24	3.65	3.31	3.08	2.92	2.80	2.71	2.63
28	5.61	4.22	3.63	3.29	3.06	2.90	2.78	2.69	2.61
29	5.59	4.20	3.61	3.27	3.04	2.88	2.76	2.67	2.59
30	5.57	4.18	3.59	3.25	3.03	2.87	2.75	2.65	2.57
40	5.42	4.05	3.46	3.13	2.90	2.74	2.62	2.53	2.45
60	5.29	3.93	3.34	3.01	2.79	2.63	2.51	2.41	2.33
120	5.15	3.80	3.23	2.89	2.67	2.52	2.39	2.30	2.22
∞	5.02	3.69	3.12	2.79	2.57	2.41	2.29	2.19	2.11

Numerator, df_1									
10	**12**	**15**	**20**	**24**	**30**	**40**	**60**	**120**	**∞**
2.70	2.60	2.50	2.39	2.33	2.27	2.21	2.14	2.08	2.00
2.67	2.57	2.47	2.36	2.30	2.24	2.18	2.11	2.04	1.97
2.64	2.54	2.44	2.33	2.27	2.21	2.15	2.08	2.01	1.94
2.61	2.51	2.41	2.30	2.24	2.18	2.12	2.05	1.98	1.91
2.59	2.49	2.39	2.28	2.22	2.16	2.09	2.03	1.95	1.88
2.57	2.47	2.36	2.25	2.19	2.13	2.07	2.00	1.93	1.85
2.55	2.45	2.34	2.23	2.17	2.11	2.05	1.98	1.91	1.83
2.53	2.43	2.32	2.21	2.15	2.09	2.03	1.96	1.89	1.81
2.51	2.41	2.31	2.20	2.14	2.07	2.01	1.94	1.87	1.79
2.39	2.29	2.18	2.07	2.01	1.94	1.88	1.80	1.72	1.64
2.27	2.17	2.06	1.94	1.88	1.82	1.74	1.67	1.58	1.48
2.16	2.05	1.94	1.82	1.76	1.69	1.61	1.53	1.43	1.31
2.05	1.94	1.83	1.71	1.64	1.57	1.48	1.39	1.27	1.00

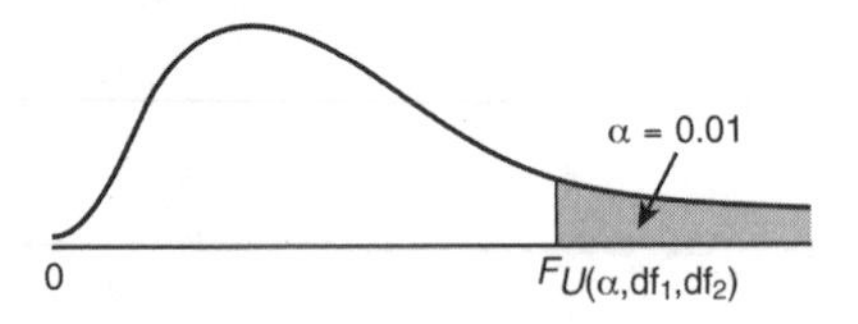

TABLE C.4 Continued

For a particular combination of numerator and denominator degrees of freedom, entry represents the critical values of F corresponding to a specified upper-tail area (α).

Denominator, df_2	Numerator, df_1 1	2	3	4	5	6	7	8	9
1	4,052.00	4,999.50	5,403.00	5,625.00	5,764.00	5,859.00	5,928.00	5,982.00	6,022.00
2	98.50	99.00	99.17	99.25	99.30	99.33	99.36	99.37	99.39
3	34.12	30.82	29.46	28.71	28.24	27.91	27.67	27.49	27.35
4	21.20	18.00	16.69	15.98	15.52	15.21	14.98	14.80	14.66
5	16.26	13.27	12.06	11.39	10.97	10.67	10.46	10.29	10.16
6	13.75	10.92	9.78	9.15	8.75	8.47	8.26	8.10	7.98
7	12.25	9.55	8.45	7.85	7.46	7.19	6.99	6.84	6.72
8	11.26	8.65	7.59	7.01	6.63	6.37	6.18	6.03	5.91
9	10.56	8.02	6.99	6.42	6.06	5.80	5.61	5.47	5.35
10	10.04	7.56	6.55	5.99	5.64	5.39	5.20	5.06	4.94
11	9.65	7.21	6.22	5.67	5.32	5.07	4.89	4.74	4.63
12	9.33	6.93	5.95	5.41	5.06	4.82	4.64	4.50	4.39
13	9.07	6.70	5.74	5.21	4.86	4.62	4.44	4.30	4.19
14	8.86	6.51	5.56	5.04	4.69	4.46	4.28	4.14	4.03
15	8.68	6.36	5.42	4.89	4.56	4.32	4.14	4.00	3.89
16	8.53	6.23	5.29	4.77	4.44	4.20	4.03	3.89	3.78
17	8.40	6.11	5.18	4.67	4.34	4.10	3.93	3.79	3.68
18	8.29	6.01	5.09	4.58	4.25	4.01	3.84	3.71	3.60
19	8.18	5.93	5.01	4.50	4.17	3.94	3.77	3.63	3.52
20	8.10	5.85	4.94	4.43	4.10	3.87	3.70	3.56	3.46
21	8.02	5.78	4.87	4.37	4.04	3.81	3.64	3.51	3.40

Numerator, df_1									
10	12	15	20	24	30	40	60	120	∞
6,056.00	6,106.00	6,157.00	6,209.00	6,235.00	6,261.00	6,287.00	6,313.00	6,339.00	6,366.00
99.40	99.42	99.43	94.45	99.46	99.47	99.47	99.48	99.49	99.50
27.23	27.05	26.87	26.69	26.60	26.50	26.41	26.32	26.22	26.13
14.55	14.37	14.20	14.02	13.93	13.84	13.75	13.65	13.56	13.46
10.05	9.89	9.72	9.55	9.47	9.38	9.29	9.20	9.11	9.02
7.87	7.72	7.56	7.40	7.31	7.23	7.14	7.06	6.97	6.88
6.62	6.47	6.31	6.16	6.07	5.99	5.91	5.82	5.74	5.65
5.81	5.67	5.52	5.36	5.28	5.20	5.12	5.03	4.95	4.86
5.26	5.11	4.96	4.81	4.73	4.65	4.57	4.48	4.40	4.31
4.85	4.71	4.56	4.41	4.33	4.25	4.17	4.08	4.00	3.91
4.54	4.40	4.25	4.10	4.02	3.94	3.86	3.78	3.69	3.60
4.30	4.16	4.01	3.86	3.78	3.70	3.62	3.54	3.45	3.36
4.10	3.96	3.82	3.66	3.59	3.51	3.43	3.34	3.25	3.17
3.94	3.80	3.66	3.51	3.43	3.35	3.27	3.18	3.09	3.00
3.80	3.67	3.52	3.37	3.29	3.21	3.13	3.05	2.96	2.87
3.69	3.55	3.41	3.26	3.18	3.10	3.02	2.93	2.81	2.75
3.59	3.46	3.31	3.16	3.08	3.00	2.92	2.83	2.75	2.65
3.51	3.37	3.23	3.08	3.00	2.92	2.84	2.75	2.66	2.57
3.43	3.30	3.15	3.00	2.92	2.84	2.76	2.67	2.58	2.49
3.37	3.23	3.09	2.94	2.86	2.78	2.69	2.61	2.52	2.42
3.31	3.17	3.03	2.88	2.80	2.72	2.64	2.55	2.46	2.36

(*continues*)

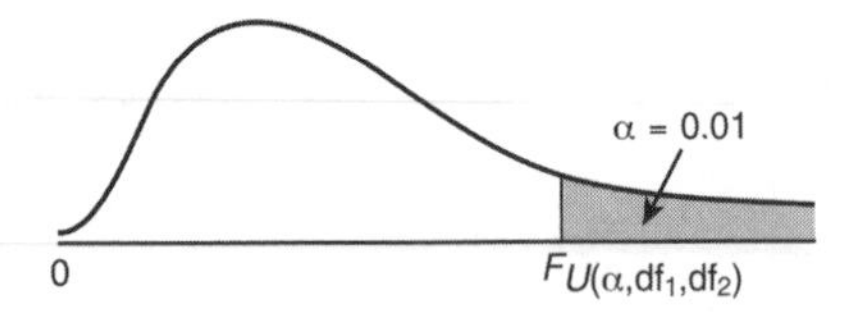

TABLE C.4 Continued

For a particular combination of numerator and denominator degrees of freedom, entry represents the critical values of F corresponding to a specified upper-tail area (α).

Denominator, df_2	Numerator, df_1 1	2	3	4	5	6	7	8	9
22	7.95	5.72	4.82	4.31	3.99	3.76	3.59	3.45	3.35
23	7.88	5.66	4.76	4.26	3.94	3.71	3.54	3.41	3.30
24	7.82	5.61	4.72	4.22	3.90	3.67	3.50	3.36	3.26
25	7.77	5.57	4.68	4.18	3.85	3.63	3.46	3.32	3.22
26	7.72	5.53	4.64	4.14	3.82	3.59	3.42	3.29	3.18
27	7.68	5.49	4.60	4.11	3.78	3.56	3.39	3.26	3.15
28	7.64	5.45	4.57	4.07	3.75	3.53	3.36	3.23	3.12
29	7.60	5.42	4.54	4.04	3.73	3.50	3.33	3.20	3.09
30	7.56	5.39	4.51	4.02	3.70	3.47	3.30	3.17	3.07
40	7.31	5.18	4.31	3.83	3.51	3.29	3.12	2.99	2.89
60	7.08	4.98	4.13	3.65	3.34	3.12	2.95	2.82	2.72
120	6.85	4.79	3.95	3.48	3.17	2.96	2.79	2.66	2.56
∞	6.63	4.61	3.78	3.32	3.02	2.80	2.64	2.51	2.41

Numerator, df_1									
10	12	15	20	24	30	40	60	120	∞
3.26	3.12	2.98	2.83	2.75	2.67	2.58	2.50	2.40	2.31
3.21	3.07	2.93	2.78	2.70	2.62	2.54	2.45	2.35	2.26
3.17	3.03	2.89	2.74	2.66	2.58	2.49	2.40	2.31	2.21
3.13	2.99	2.85	2.70	2.62	2.54	2.45	2.36	2.27	2.17
3.09	2.96	2.81	2.66	2.58	2.50	2.42	2.33	2.23	2.13
3.06	2.93	2.78	2.63	2.55	2.47	2.38	2.29	2.20	2.10
3.03	2.90	2.75	2.60	2.52	2.44	2.35	2.26	2.17	2.06
3.00	2.87	2.73	2.57	2.49	2.41	2.33	2.23	2.14	2.03
2.98	2.84	2.70	2.55	2.47	2.39	2.30	2.21	2.11	2.01
2.80	2.66	2.52	2.37	2.29	2.20	2.11	2.02	1.92	1.80
2.63	2.50	2.35	2.20	2.12	2.03	1.94	1.84	1.73	1.60
2.47	2.34	2.19	2.03	1.95	1.86	1.76	1.66	1.53	1.38
2.32	2.18	2.04	1.88	1.79	1.70	1.59	1.47	1.32	1.00

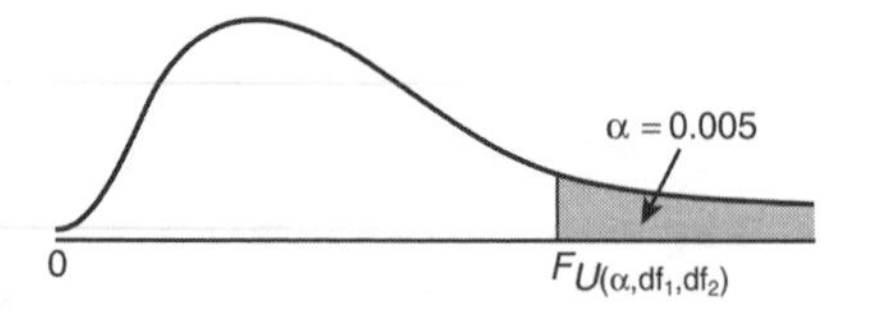

TABLE C.4 Continued

Denominator, df_2	Numerator, df_1 1	2	3	4	5	6	7	8	9
1	16,211.00	20,000.000	21,615.00	22,500.00	23,056.00	23,437.00	23,715.00	23,925.00	24,091.00
2	198.50	199.00	199.20	199.20	199.30	199.30	199.40	199.40	199.40
3	55.55	49.80	47.47	46.19	45.39	44.84	44.43	44.13	43.88
4	31.33	26.28	24.26	23.15	22.46	21.97	21.62	21.35	21.14
5	22.78	18.31	16.53	15.56	14.94	14.51	14.20	13.96	13.77
6	18.63	14.54	12.92	12.03	11.46	11.07	10.79	10.57	10.39
7	16.24	12.40	10.88	10.05	9.52	9.16	8.89	8.68	8.51
8	14.69	11.04	9.60	8.81	8.30	7.95	7.69	7.50	7.34
9	13.61	10.11	8.72	7.96	7.47	7.13	6.88	6.69	6.54
10	12.83	9.43	8.08	7.34	6.87	6.54	6.30	6.12	5.97
11	12.23	8.91	7.60	6.88	6.42	6.10	5.86	5.68	5.54
12	11.75	8.51	7.23	6.52	6.07	5.76	5.52	5.35	5.20
13	11.37	8.19	6.93	6.23	5.79	5.48	5.25	5.08	4.94
14	11.06	7.92	6.68	6.00	5.56	5.26	5.03	4.86	4.72
15	10.80	7.70	6.48	5.80	5.37	5.07	4.85	4.67	4.54
16	10.58	7.51	6.30	5.64	5.21	4.91	4.69	4.52	4.38
17	10.38	7.35	6.16	5.50	5.07	4.78	4.56	4.39	4.25
18	10.22	7.21	6.03	5.37	4.96	4.66	4.44	4.28	4.14
19	10.07	7.09	5.92	5.27	4.85	4.56	4.34	4.18	4.04
20	9.94	6.99	5.82	5.17	4.76	4.47	4.26	4.09	3.96
21	9.83	6.89	5.73	5.09	4.68	4.39	4.18	4.02	3.88
22	9.73	6.81	5.65	5.02	4.61	4.32	4.11	3.94	3.81
23	9.63	6.73	5.58	4.95	4.54	4.26	4.05	3.88	3.75
24	9.55	6.66	5.52	4.89	4.49	4.20	3.99	3.83	3.69

Numerator, df_1									
10	**12**	**15**	**20**	**24**	**30**	**40**	**60**	**120**	**∞**
24,224.00	24,426.00	24,630.00	24,836.00	24,910.00	25,044.00	25,148.00	25,253.00	25,359.00	25,465.00
199.40	199.40	199.40	199.40	199.50	199.50	199.50	199.50	199.50	199.50
43.69	43.39	43.08	42.78	42.62	42.47	42.31	42.15	41.99	41.83
20.97	20.70	20.44	20.17	20.03	19.89	19.75	19.61	19.47	19.32
13.62	13.38	13.15	12.90	12.78	12.66	12.53	12.40	12.27	12.11
10.25	10.03	9.81	9.59	9.47	9.36	9.24	9.12	9.00	8.88
8.38	8.18	7.97	7.75	7.65	7.53	7.42	7.31	7.19	7.08
7.21	7.01	6.81	6.61	6.50	6.40	6.29	6.18	6.06	5.95
6.42	6.23	6.03	5.83	5.73	5.62	5.52	5.41	5.30	5.19
5.85	5.66	5.47	5.27	5.17	5.07	4.97	4.86	4.75	1.61
5.42	5.24	5.05	4.86	4.75	4.65	4.55	4.44	4.34	4.23
5.09	4.91	4.72	4.53	4.43	4.33	4.23	4.12	4.01	3.90
4.82	4.64	4.46	4.27	4.17	4.07	3.97	3.87	3.76	3.65
4.60	4.43	4.25	4.06	3.96	3.86	3.76	3.66	3.55	3.41
4.42	4.25	4.07	3.88	3.79	3.69	3.58	3.48	3.37	3.26
4.27	4.10	3.92	3.73	3.64	3.54	3.44	3.33	3.22	3.11
4.14	3.97	3.79	3.61	3.51	3.41	3.31	3.21	3.10	2.98
4.03	3.86	3.68	3.50	3.40	3.30	3.20	3.10	2.89	2.87
3.93	3.76	3.59	3.40	3.31	3.21	3.11	3.00	2.89	2.78
3.85	3.68	3.50	3.32	3.22	3.12	3.02	2.92	2.81	2.69
3.77	3.60	3.43	3.24	3.15	3.05	2.95	2.84	2.73	2.61
3.70	3.54	3.36	3.18	3.08	2.98	2.88	2.77	2.66	2.55
3.64	3.47	3.30	3.12	3.02	2.92	2.82	2.71	2.60	2.48
3.59	3.42	3.25	3.06	2.97	2.87	2.77	2.66	2.55	2.43

(continues)

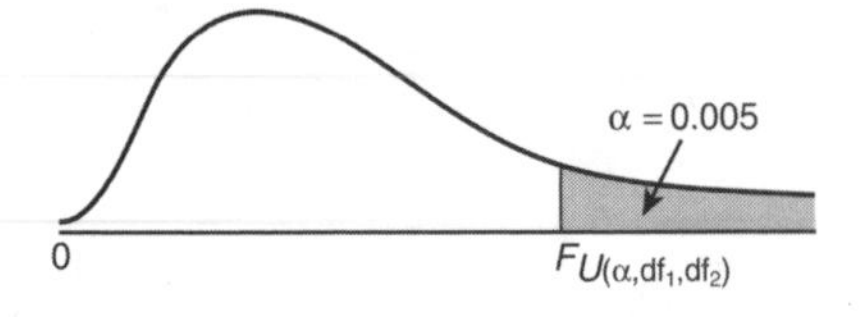

TABLE C.4 Continued

Denominator, df_2	Numerator, df_1 1	2	3	4	5	6	7	8	9
25	9.48	6.60	5.46	4.84	4.43	4.15	3.94	3.78	3.64
26	9.41	6.54	5.41	4.79	4.38	4.10	3.89	3.73	3.60
27	9.34	6.49	5.36	4.74	4.34	4.06	3.85	3.69	3.56
28	9.28	6.44	5.32	4.70	4.30	4.02	3.81	3.65	3.52
29	9.23	6.40	5.28	4.66	4.26	3.98	3.77	3.61	3.48
30	9.18	6.35	5.24	4.62	4.23	3.95	3.74	3.58	3.45
40	8.83	6.07	4.98	4.37	3.99	3.71	3.51	3.35	3.22
60	8.49	5.79	4.73	4.14	3.76	3.49	3.29	3.13	3.01
120	8.18	5.54	4.50	3.92	3.55	3.28	3.09	2.93	2.81
∞	7.88	5.30	4.28	3.72	3.35	3.09	2.90	2.74	2.62

Numerator, df_1									
10	12	15	20	24	30	40	60	120	∞
3.54	3.37	3.20	3.01	2.92	2.82	2.72	2.61	2.50	2.38
3.49	3.33	3.15	2.97	2.87	2.77	2.67	2.56	2.45	2.33
3.45	3.28	3.11	2.93	2.83	2.73	2.63	2.52	2.41	2.29
3.41	3.25	3.07	2.89	2.79	2.69	2.59	2.48	2.37	2.25
3.38	3.21	3.04	2.86	2.76	2.66	2.56	2.45	2.33	2.21
3.34	3.18	3.01	2.82	2.73	2.63	2.52	2.42	2.30	2.18
3.12	2.95	2.78	2.60	2.50	2.40	2.30	2.18	2.06	1.93
2.90	2.74	2.57	2.39	2.29	2.19	2.08	1.96	1.83	1.69
2.71	2.54	2.37	2.19	2.09	1.98	1.87	1.75	1.61	1.43
2.52	2.36	2.19	2.00	1.90	1.79	1.67	1.53	1.36	1.00

APPENDIX D

Spreadsheet Tips

These Spreadsheet Tips complement the Spreadsheet Solutions that appear throughout this book. Use these tips about creating charts and using worksheet statistical functions as a starting point to learn more about how to use Microsoft Excel to obtain your own customized results.

CT: Chart Tips

CT1 Arranging Data in Categorical Charts

You can sort your data to display categories in largest to smallest order. For bar charts, arrange the summary table in largest to smallest order. For pie charts, arrange the summary table in largest to smallest order in order to show largest to smallest pie slices in clockwise order.

CT2 Reformatting Charts

Right-click on chart components to reformat chart components. Use this technique to eliminate unwarranted gridlines and legends, change color schemes, or to change the text font and size of titles and axis labels.

CT3 Creating Charts

To create a chart, you first select the data to be charted by dragging your mouse over the data. Then you make a choice in the **Charts** group of the

Insert tab (of the **Office Ribbon**). You can restyle charts using the choices on the **Chart Tools Design** tab. (In Excel 2007, you also use the **Layout** tab of the **Chart Tools**.)

CT4 Creating Pareto Charts

To create a Pareto chart, first create a summary table with columns for categories, percentage, and cumulative percentage, arranged from largest to smallest percentage. Then create a new chart that uses this summary table. Select **Clustered Column**. In Excel 2007 and Excel 2010, you must click on the second series (the cumulative percentages) and then click **Change Chart Type** in the **Chart Tools Design** tab to change that series from a column to line type. In more recent Excels, click on the chart and then click **Change Chart Type** in the **Chart Tools Design** tab. In the Change Chart Type dialog box, click **Combo** in the left pane and select **Line with Markers** as the chart type for the second series.

Add data labels using your categories column. Adjust the primary (left) *Y*-axis so that it shows a scale from 0 to 100%. Add a secondary (right) *Y*-axis with the same scale. Reformat the chart as necessary using Tip CT2.

CT5 Creating Histograms

If you have previously chosen to use the Analysis ToolPak **Histogram** procedure, complete all the instructions in Tip ATT1 in Appendix E to create a histogram. If you have created a frequency distribution using the FREQUENCY function or have data already summarized into a frequency distribution, create a new **Column** chart (see Tip CT3, if necessary) to create a histogram. You also can create a column of midpoints to serve as the *X*-axis labels, as was done for the Fan Cost Index histogram shown in Chapter 2 and which also appears in the **Chapter 2 Histogram** file.

CT6 Creating Time-Series Plots

First, arrange your data values so that the time values to be plotted appear in the column to the immediate left of the values to be plotted on the *Y* axis. Then create a new chart and select **XY (Scatter)** as the chart type.

CT7 Creating Scatter Plots

First, arrange your data values so that the values to be plotted on the *X* (horizontal) axis appear in the column to the immediate left of the values to be plotted on the *Y* axis. Then create a new chart and select **XY (Scatter)** as the chart type.

FT: Function Tips

FT1 How to Enter Functions for Numerical Descriptive Measures

Enter the functions for numerical descriptive measures in the form ***FUNCTION(cell range of the data values)***. Use the AVERAGE (for the mean), MEDIAN, and MODE functions to calculate these measures of central tendency. Use the VAR.S (sample variance), STDEV.S (sample standard deviation), VAR.P (population variance), and STDEV.P (population standard deviation) functions to calculate measures of variation. Use the difference of the MAX (maximum value) and MIN (minimum value) functions to calculate the range.

FT2 Functions for Normal Probabilities

Use the **STANDARDIZE**, **NORM.DIST**, **NORM.S.INV**, and **NORM.INV** functions to calculate values associated with normal probabilities. Enter these functions as

- **STANDARDIZE(*X*, *mean*, *standard deviation*)**, where *X* is the *X* value of interest, and ***mean*** and ***standard deviation*** are the mean and standard deviation for a variable of interest
- **NORM.DIST(*X*, *mean*, *standard deviation*, *True*)**
- **NORM.S.INV(*P<X*)**, where ***P<X*** is the area under the curve that is less than X
- **NORM.INV(*P<X*, *mean*, *standard deviation*)**

The STANDARDIZE function returns the *Z* value for a particular *X* value, mean, and standard deviation. The NORM.DIST function returns the area or probability of less than a given *X* value. The NORM.S.INV function returns the *Z* value corresponding to the probability of less than a given *X*. The NORM.INV function returns the *X* value for a given probability, mean, and standard deviation.

APPENDIX

E

Advanced Techniques

ADV: Advanced How-Tos

ADV1 Using PivotTables to Create Two-Way Cross-Classification Tables

You use PivotTables to create two-way cross-classification tables from your data values. PivotTables are worksheet areas that act as if you had entered formulas to summarize data. PivotTables give you the ability to drill down, or look at, the unsummarized data values from which the summary information is derived.

You create a PivotTable by dragging variable names into a PivotTable form using the following process:

1. Open to the worksheet that contains your unsummarized data.
2. Select **Insert** then **PivotTable**.
3. In the Create PivotTable dialog box, leave the **Select a table or range** option selected and change, if necessary, the **Table/Range** cell range to the cell range of your unsummarized data. Then select the **New Worksheet** option and click **OK**.
4. In the **PivotTable Field List** task pane, drag the label of the row variable and drop it in the **Row Labels** box. Drag a second copy of this same label and drop it in the **Σ Values** box. (This second label changes to **Count of variable name**.) Drag the label of the column variable and drop it in the **Columns Labels** box. (Figure E.1 shows a completed layout for the webpage design study example in Section 2.1.)

5. Optionally, right-click the PivotTable and click **Table Options** in the shortcut menu that appears. In the PivotTable Options dialog box, adjust the formatting and display of the PivotTable and click **OK**.

FIGURE E.1

PivotTable Field List task pane example. (Excel 2013 version shown)

ADV2 Creating Frequency Distributions

You can use the Microsoft Excel Analysis ToolPak **Histogram** procedure or the **FREQUENCY** function to create frequency distributions. (The Histogram procedure optionally creates a histogram as well.)

Using either method, you first create a column of bin values on the worksheet that contains your data values. Bin values represent the maximum value of a group. When groups are defined in the form *low value through high value*, the bin value is the *high value*. When groups are defined in the form *low value to under high value*, as they are in Chapter 2, the bin value is a number just smaller than the *high value*. For example, for the group "200 to under 250," the corresponding bin value could be 249.99, a number "just smaller" than 250.

With the column of bin values created, continue by either using the **Histogram** procedure (see Tip ATT1 later in this appendix) or the **FREQUENCY** function. If using the FREQUENCY function, select the cells to contain the frequency counts (the cell range B3:B12 in the example on the next page). Type, but do not press the **Enter** or **Tab** key, a formula in the form =**FREQUENCY(*cell range of the data values*, *cell range of the bins*)**. Then, while holding down the **Control** and **Shift** keys (or the **Command** key on a Mac), press the **Enter** key. This enters the formula as an "array formula" in all the cells you previously selected. To edit or clear the array formula, you must select all the cells, make your change, and again press **Enter** while holding down **Control** and **Shift** (or **Command**).

The example below shows columns A through C of the Histogram worksheet found in the **Chapter 2 Histogram** file. The worksheet includes formulas to calculate the total and percentage, if necessary. The ***cell range of the data values*** in the **FREQUENCY** function has been entered as Data!B2:B31 and not as B2:B31 because the data values are found on another worksheet—the Data worksheet. Using the name of a worksheet (followed by an exclamation point) as a prefix directs Microsoft Excel to that worksheet.

	A	B	C
1	Frequency Distribution for Fan Cost Index		
2	Bins	Frequency	Percentage
3	249.99	=FREQUENCY(Data!B2:B31,A3:A12)	=B3/B$13
4	299.99	=FREQUENCY(Data!B2:B31,A3:A12)	=B4/B$13
5	349.99	=FREQUENCY(Data!B2:B31,A3:A12)	=B5/B$13
6	399.99	=FREQUENCY(Data!B2:B31,A3:A12)	=B6/B$13
7	449.99	=FREQUENCY(Data!B2:B31,A3:A12)	=B7/B$13
8	499.99	=FREQUENCY(Data!B2:B31,A3:A12)	=B8/B$13
9	549.99	=FREQUENCY(Data!B2:B31,A3:A12)	=B9/B$13
10	599.99	=FREQUENCY(Data!B2:B31,A3:A12)	=B10/B$13
11	649.99	=FREQUENCY(Data!B2:B31,A3:A12)	=B11/B$13
12	699.99	=FREQUENCY(Data!B2:B31,A3:A12)	=B12/B$13
13	Total	=SUM(B3:B12)	=SUM(C3:C12)

ADV3 Using the Ampersand Operator to Form Labels

The COMPUTE worksheet of **Chapter 5 Normal** uses formulas in columns A and D to dynamically create labels based on the data values you enter. These formulas make extensive use of the ampersand operator (&) to construct the actual label. For example, the cell A10 formula = "P(X< = "&B8&")" results in the display of **P(X< = 6.2)** because the contents of cell B8, 6.2 are combined with "P(X< = "and")". If you entered the value 9 in cell B8, the label in cell A10 would change to **P(X< = 9)**. Open the Bonus Sheet worksheet of **Chapter 5 Normal** to see all COMPUTE worksheet formulas that use the ampersand operator.

ADV4 Modifying Chapter 6 Sigma Unknown for Use with Unsummarized Data

To modify **Chapter 6 Sigma Unknown** for use with unsummarized data, first enter the unsummarized data in column E. Enter a variable label for the data in cell E1 and then enter values in the cells of the subsequent rows. Do not skip a row.

Next change the entries in cells B4 through B6. Select cell B4, enter =STDEV.S(E:E), and press **Enter**. Select cell B5, enter =AVERAGE(E:E), and press **Enter**. Select cell B6, enter =COUNT(E:E), and press **Enter**. The worksheet updates all calculations and displays the confidence interval estimate for the unsummarized data.

ADV5 Modifying Chapter 8 Paired T for Use with Other Data Sets

How you modify **Chapter 8 Paired T** for use with another data set depends on the number of values found in that other data set. If the data set contains exactly 15 values, enter the data into the worksheet using the instructions found in the worksheet. Otherwise, select cell range E2:H2.

If the data set contains less than 15 values, first select the cell range E3:H3. Then follow these steps:

- Right-click over the selected cell range and click **Delete** in the shortcut menu that appears.
- In the Delete dialog box, click **Shift Cells Up** and click **OK**.

Repeat the two bulleted steps as many times as is necessary to shorten the data rows to the number of values in the data set. Then enter the new data in columns E through G.

If the data contains more than 15 values, follow these steps:

- Right-click over the selected cell range and click **Insert** in the shortcut menu that appears.
- In the Insert dialog box, click **Shift cells down** and click **OK**.

Repeat the two bulleted steps as many times as is necessary to lengthen the data rows to the number of values in the data set. Then select cell H2 and copy its formula down to the new blank cells in column H that were just inserted. As a last step, enter the new data in columns E through G.

ADV6 Using the LINEST Function to Calculate Regression Results

You can use the **LINEST** function to calculate regression results in Microsoft Excel. These results are similar to those calculated by the Analysis ToolPak **Regression** procedure.

To use this function, first, with your mouse, select an empty cell range that is five rows deep and contains the number of columns that is equal to the number of your independent *X* variables plus one. For a simple linear regression model, select five rows by two columns; for a multiple regression model with two independent variables, select five rows by three columns.

Type, but do not press the **Enter** or **Tab** key, a formula in the form =LINEST(*cell range of the dependent variable*, *cell range of the independent variables*, **True**, **True**). Then, while holding down the **Control** and **Shift** keys, press the **Enter** key. This enters the formula as an "array formula" in all the cells you previously selected. To edit or clear the formula, you must select all the cells, make your change, and again press **Enter** while holding down **Control** and **Shift**. (If you are using an Apple keyboard, you can hold down **Command** instead of holding down **Control** and **Shift** for this operation.)

The results returned by the array formula are unlabeled. For a multiple regression model, some results appear as **#N/A**, which is not an error. Add labels in the columns immediately to the left and right of the results area to label the results, as shown in the example below.

In **Appendix E Regression**, the **SLR_LINEST** and the **MR_LINEST** worksheets illustrate this labeling technique. Shown below is the SLR_LINEST worksheet using the moving company study data of Chapter 10. In this worksheet, the LINEST array function has been entered into the cell range B3 through C7.

	A	B	C	D
1	Regression Analysis for Moving Company Study			
2				
3	Cubic Feet Moved *Coefficient* (b_1)	0.0501	-2.3697	Intercept *Coefficent* (b_0)
4	Cubic Feet Moved *Standard Error*	0.0030	2.0733	Intercept *Standard Error*
5	R Square	0.8892	5.0314	Standard Error
6	*F*	272.9864	34	Residual *df*
7	Regression *MS*	6910.7189	860.7186	Residual *MS*

ADV7 Modifying Chapter 10 Simple Linear Regression for Use with Other Data Sets

To modify **Chapter 10 Simple Linear Regression** for use with other data sets, open to the SLRData worksheet and paste the new X variable data into column A and the new Y variable data in column B, replacing the data already there. Then open to the SLR worksheet and select the cell range L2:M6. Edit the cell range to reflect the length of the new data set and then, while holding the **Control** and **Shift** keys (or the **Command** key on a Mac), press Enter. The worksheet updates the results to reflect the new data set.

Modifying the RESIDUALS worksheet is a multistep process. First paste the X variable data into column B and the Y variable data in column D of the RESIDUALS worksheet. Then, for sample sizes smaller than 36, delete the extra rows. For sample sizes greater than 36, copy the column C and E formulas down through the row containing the last pair of X and Y values, and add the new observation numbers in column A.

ATT: Analysis ToolPak Tips

The Analysis ToolPak add-in component of Microsoft Excel adds a number of statistical procedures. Use the instructions in Section A.S2 to verify that this add-in is installed on your computer. (The Analysis ToolPak is *not* usually installed when you initially set up Microsoft Excel and is not included with Excel 2008 for Macs.)

To use an Analysis ToolPak procedure, first select **Data Analysis** from the **Data** tab. In the Data Analysis dialog box that appears (shown below), click a procedure name and then click the **OK** button. For most procedures, a second dialog box appears, in which you make entries and selections to complete the procedure.

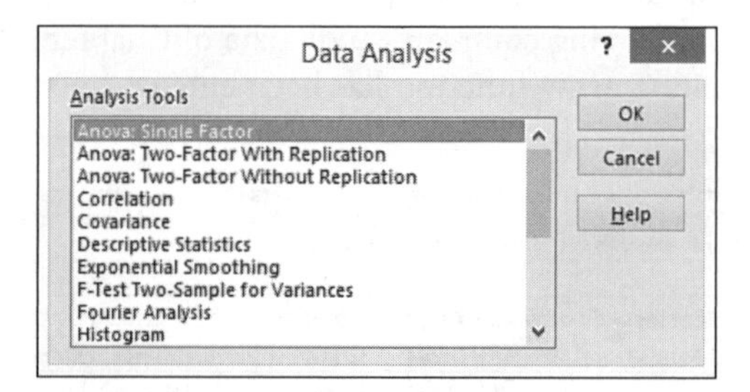

ATT1 Histogram Procedure

Begin by making sure that the worksheet that contains your data values also contains a column of bin values (see Tip ADV2). In the Histogram dialog box, enter the cell ranges of the data values and bin values and check **Labels** if these ranges begin with a column heading. Check **Chart Output** if you want to create a histogram as well as the default frequency distribution.

ATT2 Descriptive Statistics Procedure

Place the data values for the variable to be summarized in a column, using the row 1 cells for column labels. In the Descriptive Statistics dialog box, enter that column range as the **Input Range**, click **Columns** in the **Grouped By** set, and check **Labels in First Row**. Then click **New Worksheet Ply** and check **Summary statistics**. A table of descriptive statistics appears on a new worksheet.

ATT3 t-Test: Two-Sample Assuming Equal Variances Procedure

Place the data for the two groups in separate columns, using the row 1 cells for column (group) labels. In the t-Test: Two-Sample Assuming Equal Variances dialog box, enter the group 1 cell range as the **Variable 1 Range** and enter the group 2 cell range as the **Variable 2 Range**. Enter **0** as the **Hypothesized Mean Difference** and check **Labels.** Enter **0.05** as the **Alpha** value, select the **New Worksheet Ply** option, and click **OK**. The results appear on a new worksheet similar to the one shown here.

	A	B	C
1	t-Test: Two-Sample Assuming Equal Variances		
2			
3		*City*	*Suburban*
4	Mean	49.3	44.4
5	Variance	222.4592	129.4694
6	Observations	50	50
7	Pooled Variance	175.9643	
8	Hypothesized Mean Difference	0	
9	df	98	
10	t Stat	1.84694	
11	P(T<=t) one-tail	0.03389	
12	t Critical one-tail	1.66055	
13	P(T<=t) two-tail	0.06777	
14	t Critical two-tail	1.98447	

ATT4 t-Test: Paired Two Sample for Means Procedure

Place the data for the two groups in separate columns, using the row 1 cells for column (group) labels. In the t-Test: Paired Two Sample for Means dialog box, enter the group 1 sample data cell range as the **Variable 1 Range** and enter the group 2 sample data cell range as the **Variable 2 Range**. Enter **0** as the **Hypothesized Mean Difference** and check **Labels.** Enter **0.05** as the **Alpha** value, select the **New Worksheet Ply** option, and click **OK**. The results appear on a new worksheet similar to the one shown here.

	A	B	C
1	t-Test: Paired Two Sample for Means		
2			
3		*TV*	*Internet*
4	Mean	64.6429	68.0714
5	Variance	35.4780	25.4560
6	Observations	14	14
7	Pearson Correlation	0.9608	
8	Hypothesized Mean Difference	0	
9	df	13	
10	t Stat	-7.1862	
11	P(T<=t) one-tail	0.0000	
12	t Critical one-tail	1.7709	
13	P(T<=t) two-tail	0.0000	
14	t Critical two-tail	2.1604	

ATT5 ANOVA: Single Factor Procedure

Place the data of each group in its own column, using the row 1 cells for column (group) labels. In the ANOVA: Single Factor dialog box, enter the cell range for *all* of your data as the **Input Range**. Select **Columns** and check **Labels in First Row**. Enter **0.05** as the **Alpha** value, select the **New Worksheet Ply** option, and click **OK**. The results appear on a new worksheet that will be similar to the Chapter 9 one-way ANOVA spreadsheets except that there will be a row that contains the level of significance. **Chapter 9 ATP One-Way ANOVA** shows the Analysis ToolPak results for the Chapter 9 WORKED-OUT PROBLEM 7.

ATT6 Regression Procedure

Place the data for each variable in its own column, using the row 1 cells for column (group) labels. Use the first column for the dependent variable *Y* and use the second and subsequent columns for your independent *X* variables. (Simple linear regression, discussed in Chapter 10, uses only one independent variable.) In the Regression dialog box, enter the cell range for dependent variable *Y* as the **Input Y Range** and enter the cell range for independent *X* variable or variables as the **Input X Range**. Check **Labels** and **Confidence Level** and enter **95** in the percentage box. To assist in a residual analysis, also check **Residuals** and **Residual Plots**. Click **OK**. The results appear on a new worksheet that contains a layout similar to the regression results spreadsheets shown in Chapters 10 and 11.

Chapter 10 ATP Simple Linear Regression contains the ToolPak regression results for the Chapter 10 WORKED-OUT PROBLEM 2 moving company study data. **Chapter 11 ATP Multiple Regression** contains the ToolPak regression results for the Chapter 11 WORKED-OUT PROBLEM 1 moving company study data.

APPENDIX F

Documentation for Downloadable Files

This appendix lists and describes all of the files that you can download for free at **www.ftpress.com/evenyoucanlearnstatistics3e** for use with this book.

F.1 Downloadable Data Files

Throughout this book, the file icon identifies downloadable data files that allow you to examine the data for selected problems.

Each data file is available as:

- An Excel .xlsx workbook file
- Either a TI.83m matrix file or a TI .83l list file that can be loaded into any calculator in the TI-83 and TI-84 family

The following table identifies the columns of data found in each file and identifies the chapter or chapters in which the data is used. For the Excel workbook files, the columns of data map to the lettered columns of a worksheet. For the TI .83m matrix files, the columns map to the columns of the matrix variable [D]. In the special case of a single-column data file such as **Sushi** and **Times**, the single column of data is stored as a TI .83l list file.

Column names that appear in italics, such as *Team* in the **Baseball** file, are row label columns and are not included in the TI matrix and list files.

Advertise	Sales, radio advertising expenses, televison advertising expenses (Chapter 14)
Anscombe	Data sets A, B, C, and D each with 11 pairs of *X* and *Y* values (Chapter 10)
Auto	Miles per gallon, horsepower, and weight for a sample of 50 car models (Chapter 11)
Baseball	*Team*, earned run average, runs scored per game, wins (Chapters 10 and 11)
BottledWater	*Year*, per capita consumption of botted water (Chapter 2)
BoxFills	Plant 1 cereal weights, Plant 2 cereal weights, Plant 3 cereal weights, and Plant 4 cereal weights (Chapter 9)
CatFood	Ounces eaten of kidney, shrimp, chicken liver, salmon, and beef (Chapter 9)
Cereals	*Cereal brand*, calories, grams of sugar (Chapter 2)
Concrete	*Sample*, compressive strength after two days, and compressive strength after seven days (Chapter 8)
DomesticBeer	*Brand*, alcohol percentage, calories, and carbohydrates (Chapters 2, 3, 5, and 6)
FastFoodChain	Burger, chicken, sandwich, and pizza market segments means sales per unit (Chapter 9)
FoodPrices	*Company*, total cost, type (Chapter 3)
GlenCove	Fair market value of house, property size in acres, and age of house (Chapters 10 and 11)
GolfBall	Design 1 distance traveled, Design 2 distance traveled, Design 3 distance traveled, Design 4 distance traveled (Chapter 9)
Insurance	Time to process applications (Chapter 6)
Intaglio	Surface hardness of untreated steel plates and surface hardness of treated steel plates (Chapter 8)
Math	Math scores using set A materials, math scores using set B materials, math scores using set C materials (Chapter 9)
Movie Revenues	*Year* and annual revenues from movies (Chapter 2)
Moving	Labor hours, cubic feet moved, and number of large pieces of furniture moved (Chapters 2, 10, and 11)
Myeloma	*Patient*, measurement before transplant, and measurement after transplant (Chapter 8)

NBACost	*Team*, fan cost index (the cost of four tickets, two beers, four soft drinks, four hot dogs, two game programs, two caps, and the parking fee for one car) (Chapters 2, 3, and 5)
OnlinePrices	*Company*, total cost (Chapter 6)
OrderTimePopulation	Population of order times from a website (Chapter 6)
Phone	Location 1 time (in minutes) to clear line problems and location 2 time to clear line problems for samples of 20 customer problems reported to each location (Chapter 8)
PropertyTaxes	*State* and property taxes per capita (Chapter 2)
Protein	*food item*, calories, protein, percentage of calories from fat, percentage of calories from saturated fat, and cholesterol (Chapter 6)
Redwood	Height, breast-height diameter, and bark thickness (Chapter 11)
Restaurants	*Location*, food rating, decor rating, service rating, summated rating, coded location (0 = urban, 1 = suburban), and price of restaurants (Chapters 3, 5, 6, 8, and 10)
SamplesofOrderTimes	Sample 1 order times from a website, sample 2 order times, sample 3 order times, and sample 20 order times (Chapter 6)
Satisfaction	Satisfaction, delivery time difference, previous (Chapter 14)
Sedans	Miles per gallon (Chapter 3)
Sports	*Sport*, movement speed, rules, team oriented, amount of contact (Chapter 14)
Supermarket	*Pair*, sales using new package, sales using old package, and difference in sales (Chapter 8)
TargetWalmart	*Item number*, Target cost, Walmart cost (Chapter 8)
Telecom	*Provider*, TV rating, Internet rating (Chapter 8)
Times	Times to get ready (Chapter 3)

Use the TI DataEditor component of TI Connect (see Section AC4 of online Appendix A) to transfer a downloaded TI data file to your calculator. Open TI Connect and click the TI DataEditor icon. When the DataEditor window opens, select **File → Open** and navigate to the file you want to transfer. Then click the **Send File** icon to transfer the contents of the data file to your calculator.

Downloadable *Spreadsheet Solution* Files

Also available for download are the Excel workbook files that are mentioned in the *Spreadsheet Solution* sections of this book. The following is a complete list of the Spreadsheet Solution Excel workbook files:

Chapter 2 Bar

Chapter 2 Histogram

Chapter 2 Pareto

Chapter 2 Pie

Chapter 2 Scatter Plot

Chapter 2 Time-Series

Chapter 2 Two-Way PivotTable

Chapter 2 Two-Way

Chapter 3 BoxWhisker Plot

Chapter 3 Descriptive

Chapter 5 Binomial

Chapter 5 Normal

Chapter 5 Poisson

Chapter 6 Proportion

Chapter 6 Sigma Unknown

Chapter 8 ATP Paired T

Chapter 8 ATP Pooled-Variance T

Chapter 8 Paired T

Chapter 8 Pooled-Variance T with Sample Statistics

Chapter 8 Pooled-Variance T with Unsummarized Data

Chapter 8 Z Two Proportions

Chapter 9 ATP One-Way ANOVA

Chapter 9 Chi-Square

Chapter 9 Chi-Square Spreadsheets

Chapter 9 One-Way ANOVA

Chapter 10 ATP Simple Linear Regression

Chapter 10 Simple Linear Regression

Chapter 11 Multiple Regression

Chapter 11 ATP Multiple Regression

Chapter 13 Bullet Graph

Chapter 13 Drilldown

Chapter 13 Sparklines

Chapter 13 Treemap

Appendix E Regression

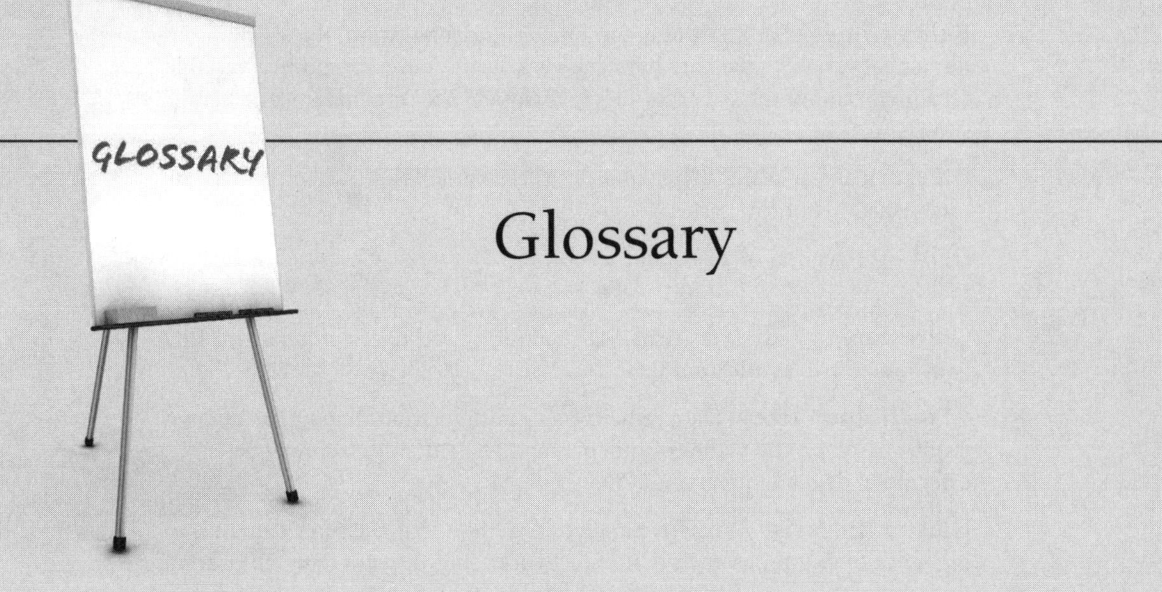

Glossary

Alternative hypothesis (H_1)—The opposite of the null hypothesis (H_0).

Analysis of variance (ANOVA)—A statistical method that tests the effect of different factors on a variable of interest.

Analytics—Descriptive and inferential methods that are focused on discovering patterns to or generating hypotheses about data already collected.

Bar chart—A chart containing rectangles ("bars") in which the length of each bar represents the count, amount, or percentage of responses in each category.

Big Data—High volume, high velocity, or high variety collections of data that require innovative forms of information processing for enhanced insight and decision making.

Binomial distribution—A distribution that finds the probability of a given number of successes for a given probability of success and sample size.

Box-and-whisker plot—Also known as a **boxplot**; a graphical representation of the five-number summary that consists of the smallest value, the first quartile (or 25th percentile), the median, the third quartile (or 75th percentile), and the largest value.

Bullet Graph—Bar chart that combines the visualization of a single numerical variable (the bar) with the visualization of a categorical variable (shadings behind the bar) that helps to classify the numerical value being shown.

Categorical variable—The values of these variables are selected from an established list of categories.

Cell—Intersection of a row and a column in a two-way cross-classification table.

Chi-square (χ^2) distribution—Distribution used to test relationships in two-way cross-classification tables.

Classification tree—The predictive analytics method that splits data into groups based on the values of independent (X) variables to predict a categorical dependent variable (Y).

Cluster Analysis—Predictive analytics method that classifies data into a sequence of groupings such that objects of each group are more alike among themselves than they are to objects found in other groups.

Coefficient of correlation—Measures the strength of the linear relationship between two variables.

Coefficient of determination—Measures the proportion of variation in the dependent variable *Y* that is explained by the independent variable *X* in the regression model.

Collectively exhaustive events—One in a set of these events must occur.

Completely randomized design—Also known as **one-way ANOVA**; an experimental design in which only a single factor is used to evalvate a single factor of interest.

Confidence interval estimate—An estimate of the population parameter in the form of an interval with a lower and upper limit.

Continuous numerical variables—The values of these variables are measurements.

Critical value—Divides the nonrejection region from the rejection region.

Dashboards—Visual display that summarizes the most important variables needed for decision making or achieving a business objective.

Data Mining—Predictive analytics method that looks for previously unknown patterns in Big Data collections.

Degrees of freedom—The number of values that are free to vary.

Dependent variable—The variable to be predicted in a regression analysis. Also known as the response variable.

Descriptive analytics—Methods that summarize large collections of data or variables to present constantly updated status information.

Descriptive statistics—The branch of statistics that focuses on collecting, summarizing, and presenting a set of data.

Discrete numerical variables—The values of these variables are counts of things.

Drill-down—The revealing of the data that underlies a higher-level summary.

Error sum of squares (*SSE*)—Consists of variation that is due to factors other than the relationship between X and Y in a regression analysis.

Event—Each possible type of occurrence.

Expected frequency—Frequency expected in a particular cell of a cross-classification table if the null hypothesis is true.

Expected value—The mean of a probability distribution.

Experiments—A process that uses controlled conditions to study the effect on the variable of interest of varying the value(s) of another variable or variables.

Explanatory variable—A variable used to predict the dependent or response variable in a regression analysis. Also known as an independent variable.

***F* distribution**—A distribution used for testing the ratio of two variances in the analysis of variance and regression.

First quartile Q_1—The value such that 25.0% of the values are smaller and 75.0% are larger.

Five-number summary—Consists of smallest value, Q_1, median, Q_3, largest value.

Frame—The list of all items in the population from which samples will be selected.

Frequency distribution—A table of grouped numerical data in which the names of each group are listed in the first column and the percentages in each group are listed in the second column.

Histogram—A special bar chart for grouped numerical data in which the frequencies or percentages in each group are represented as individual bars.

Hypothesis testing—Methods used to make inferences about the hypothesized values of population parameters using sample statistics.

Independent events—Events in which the occurrence of one event in no way affects the probability of the second event.

Independent variable—A variable used to predict the dependent or response variable in a regression analysis. Also known as an explanatory variable.

Inferential statistics—The branch of statistics that analyzes sample data to reach conclusions about a population.

Joint event—An outcome that satisfies two or more criteria.

Level of significance—Probability of committing a Type I error.

Mean—The balance point in a set of data that is calculated by summing the observed numerical values in a set of data and then dividing by the number of values involved.

Mean squares—The variances in an Analysis of Variance table.

Median—The middle value in a set of data that has been ordered from the lowest to highest value.

Mode—The value in a set of data that appears most frequently.

Multidimensional scaling—Predictive analytics method that visualizes objects in a map of two or more dimensions.

Multiple regression—Regression analysis when there is more than one independent variable.

Mutually exclusive events—Events are mutually exclusive if both events cannot occur at the same time.

Normal distribution—The normal distribution is defined by its mean (μ) and standard deviation (σ) and is bell shaped.

Normal probability plot—A graphical device to evaluate whether a set of data follows a normal distribution.

Null hypothesis (H_0)—A statement about a parameter equal to a specific value, or the statement that no difference exists between the parameters for two or more populations.

Numerical variables—The values of these variables involve a count or measurement.

Observed frequency—Actual tally in a particular cell of a cross-classification table.

***p*-value**—The probability of computing a test statistic equal to or more extreme than the result found from the sample data, given that the null hypothesis H_0 is true.

Paired samples—Items are matched according to some characteristic and the differences between the matched values are analyzed.

Parameter—A measure that describes a characteristic of a population.

Pareto chart—A special type of bar chart in which the count, amount, or percentage of responses of each category are presented in descending order left to right, along with a superimposed plotted line that represents a running cumulative percentage.

Percentage distribution—A table of grouped numerical data in which the names of each group are listed in the first column and the percentages in each group are listed in the second column.

Pie chart—A circle chart in which wedge-shaped areas ("pie slices") represent the count, amount, or percentage of each category and the circle (the "pie") itself represents the total.

Poisson distribution—A distribution to find the probability of the number of occurrences in an area of opportunity.

Population—All the members of a group about which you want to draw a conclusion.

Power of a statistical test—The probability of rejecting the null hypothesis when it is false and should be rejected.

Predictive analytics—Methods that identify what is likely to occur in the near future and find relationships in data that may not be readily apparent using descriptive analytics.

Probability—The numerical value representing the chance, likelihood, or possibility a particular event will occur.

Probability distribution for a discrete random variable—A listing of all possible distinct outcomes and the probability that each will occur.

Probability sampling—A sampling process that takes into consideration the chance that each item will be selected.

Published sources—Data available in print or in electronic form, including data found on Internet websites.

Range—The difference between the largest and smallest values in a set of data.

Region of rejection—Consists of the values of the test statistic that are unlikely to occur if the null hypothesis is true.

Regression coefficients—The Y intercept and slope terms in the regression model.

Regression sum of squares (*SSR*)—Consists of variation that is due to the relationship between X and Y.

Regression tree—The predictive analytics method that splits data into groups based on the values of independent (X) variables to predict a numerical dependent variable (Y).

Residual—The difference between the observed and predicted values of the dependent variable for given values of the *X* variable(s).

Response variable—The variable to be predicted in a regression analysis. Also known as the dependent variable.

Sample—The part of the population selected for analysis.

Sampling—The process by which members of a population are selected for a sample.

Sampling distribution—The distribution of a sample statistic (such as the mean) for all possible samples of a given size *n*.

Sampling error—Variation of the sample statistic from sample to sample.

Sampling with replacement—A sampling method in which each selected item is returned to the frame from which it was selected so that it has the same probability of being selected again.

Sampling without replacement—A sampling method in which each selected item is not returned to the frame from which it was selected. Using this technique, an item can be selected only once.

Scatter plot—A chart that plots the values of two variables for each response. In a scatter plot, the *X* axis (the horizontal axis) always represents units of one variable and the *Y* axis (the vertical axis) always represents units of the second variable.

Simple linear regression—A statistical technique that uses a single numerical independent variable *X* to predict the numerical dependent variable *Y* and assumes a linear or straight-line relationship between *X* and *Y*.

Simple random sampling—The probability sampling process in which every individual or item from a population has the same chance of selection as every other individual or item.

Skewness—A skewed distribution is not symmetric. An excess of extreme values are in either the lower portion of the distribution or the upper portion of the distribution.

Slope—The change in *Y* per unit change in *X*.

Sparklines—Chart that visualizes measurements taken over time as small, compact graphs designed to appear as part of a table or written passage.

Standard deviation—Measure of variation around the mean of a set of data.

Standard error of the estimate—The standard deviation around the line of regression.

Statistic—A numerical measure that describes a characteristic of a sample.

Statistics—The branch of mathematics that consists of methods of processing and analyzing data to better support rational decision-making processes.

Sum of squares among groups (*SSA*)—The sum of the squared differences between the sample mean of each group and the mean of all the values, weighted by the sample size in each group.

Sum of squares total (*SST*)—Represents the sum of the squared differences between each individual value and the mean of all the values.

Sum of squares within groups (*SSW*)—Measures the difference between each value and the mean of its own group and sums the squares of these differences over all groups.

Summary table—A two-column table in which the names of the categories are listed in the first column and the count, amount, or percentage of responses are listed in a second column.

Survey—A data collection method that uses questionnaires or other approaches to gather responses from a set of participants.

Symmetry—Distribution in which each half of a distribution is a mirror image of the other half of the distribution.

***t* distribution**—A distribution used to develop a confidence interval estimate of the mean of a population and to test hypotheses about means and slopes.

Test statistic—The statistic used to determine whether to reject the null hypothesis.

Third quartile *Q*—The value such that 75.0% of the values are smaller and 25.0% are larger.

Time-series plot—A chart in which each point represents a response at a specific time. In a time series plot, the X axis (the horizontal axis) always represents units of time and the Y axis (the vertical axis) always represents units of the numerical responses.

Treemap—Chart that visualizes the comparison of two or more variables using the size and color of rectangles to represent values.

Two-way cross-classification table—A table that presents the count or percentage of joint responses to two categorical variables (a mutually exclusive pairing, or cross-classifying, of categories from each variable). The categories of one variable form the rows of the table, whereas the categories of the other variable form the columns.

Type I error—Occurs if the null hypothesis H_0 is rejected when it is true and should not be rejected. The probability of a Type I error occurring is α.

Type II error—Occurs if the null hypothesis H_0 is not rejected when it is false and should be rejected. The probability of a Type II error occurring is β.

Variable—A characteristic of an item or an individual that will be analyzed using statistics.

Variance—The square of the standard deviation.

Variation—The amount of dispersion, or "spread," in the data.

***Y* intercept**—The value of Y when $X = 0$.

***Z* score**—The difference between the value and the mean, divided by the standard deviation.

Index

C

D

E

F–G

H

I–J–K

L

M

Q–R

S

T–U

V–W

X–Y–Z